T0293768

Radioecology: Radiation Dynamics, Impact and Future

Radioecology: Radiation Dynamics, Impact and Future

Marty Oates

SYRAWOOD
PUBLISHING HOUSE

New York

Published by Syrawood Publishing House,
750 Third Avenue, 9th Floor,
New York, NY 10017, USA
www.syrawoodpublishinghouse.com

Radioecology: Radiation Dynamics, Impact and Future
Marty Oates

International Standard Book Number: 978-1-64740-353-9 (Hardback)

Cataloging-in-Publication Data

Radioecology : radiation dynamics, impact and future / Marty Oates.
    p. cm.
Includes bibliographical references and index.
ISBN 978-1-64740-353-9
1. Radioecology. 2. Radioactive pollution. 3. Radioactive pollution--Physiological effect. 4. Radiobiology. I. Oates, Marty.
QH543.5 .R33 2023
574.522 2--dc23

Table of Contents

Preface   IX

Chapter 1 **Understanding the Consequences
of Radiation for Ecosystem and Human
Health**   1
Manabu Fukumoto

Part I **The Impact of Fukushima Nuclear Power
Plant Accident on the Environment**   14

Chapter 2 **Fukushima Accident: Radiation Exposure to
Mice**   16
Manabu Onuma, Daiji Endoh, Hiroko Ishiniwa
and Masanori Tamaoki

Chapter 3 **FNPP Accident: Effect on Fishing Industry**   30
Takami Morita, Daisuke Ambe, Shizuho Miki,
Hideki Kaeriyama and Yuya Shigenobu

Chapter 4 **Impact of Contamination due to
Radioactive Cesium on Detritivores and
Arthropods**   41
Sota Tanaka, Tarô Adati, Tomoyuki Takahashi
and Sentaro Takahashi

Chapter 5 **FNPP Accident and Strontium Pollution in
the Environment**   51
Kazuma Koarai, Yasushi Kino, Toshitaka Oka,
Atsushi Takahashi, Toshihiko Suzuki,
Yoshinaka Shimizu, Mirei Chiba, Ken Osaka,
Keiichi Sasaki, Yusuke Urushihara,
Tomokazu Fukuda, Emiko Isogai,
Hideaki Yamashiro, Manabu Fukumoto,
Tsutomu Sekine and Hisashi Shinoda

Part II   **Consequences of Exposure to Radiation in Wild Organisms**   61

Chapter 6   **Impact on Invertebrates in the Intertidal Zone**   63
Toshihiro Horiguchi, Keita Kodama,
Gen Kume and Ik Joon Kang

Chapter 7   **mtDNA Mutations in Mano River Salmon**   87
Muhammad Fitri Bin Yusof, Gyo Kawada,
Masahiro Enomoto, Atsushi Tomiya,
Masato Watanabe, Daigo Morishita,
Shigehiko Izumi and Masamichi Nakajima

Chapter 8   **Coniferous Trees after the FNPP Accident**   97
Yoshito Watanabe

Part III   **Impact of Radiation on Livestock**   108

Chapter 9   **Female Fertilities of Domestic Animals: Impact of Low-Dose-Rate Radiation**   110
Yasuyuki Abe, Hideaki Yamashiro, Yasushi Kino,
Toshinori Oikawa, Masatoshi Suzuki,
Yusuke Urushihara, Yoshikazu Kuwahara,
Motoko Morimoto, Jin Kobayashi,
Tsutomu Sekine, Tomokazu Fukuda,
Emiko Isogai and Manabu Fukumoto

Chapter 10   **FNPP Accident and the Transgenerational Impact on a Calf**   121
Banri Suzuki, Shigefumi Tanaka,
Kohichi Nishikawa, Chikako Yoshida,
Takahisa Yamada, Yasuyuki Abe,
Tomokazu Fukuda, Jin Kobayashi,
Gohei Hayashi, Masatoshi Suzuki,
Yusuke Urushihara, Kazuma Koarai,
Yasushi Kino, Tsutomu Sekine,
Atsushi Takahasi, Toshihiro Shimizu,
Hisashi Shinoda, Kazuki Saito, Emiko Isogai,
Koh Kawasumi, Satoshi Sugimura,
Hideaki Yamashiro and Manabu Fukumoto

Chapter 11   **Pigs Immune System and the Impact of FNPP Accident**   **135**
Motoko Morimoto, Ayaka Kato, Jin Kobayashi,
Kei Okuda, Yoshikazu Kuwahara, Yasushi Kino,
Yasuyuki Abe, Tsutomu Sekine,
Tomokazu Fukuda, Emiko Isogai and
Manabu Fukumoto

Chapter 12   **A Study of Farm Animals: DNA damage due to Radiation**   **148**
Asako J. Nakamura

Part IV   **Fukushima Accident and Dose Estimation**   **157**

Chapter 13   **A Study of Cattle using Electron Spin Resonance (ESR) Tooth Dosimetry**   **159**
Kazuhiko Inoue, Ichiro Yamaguchi and
Masahiro Natsuhori

Chapter 14   **Japanese Macaques: Cumulative Dose due to External and Internal Exposures**   **172**
Satoru Endo, Kenichi Ishii, Masatoshi Suzuki,
Tsuyoshi Kajimoto, Kenichi Tanaka and
Manabu Fukumoto

Part V   **Aftermath**   **185**

Chapter 15   **A Study of Radioactive Cs in the Environment**   **187**
Kazuhiko Ninomiya

Chapter 16   **Insoluble Radioactive Cs Bearing Particles and their Physicochemical Properties**   **197**
Masatoshi Suzuki, Kazuhiko Ninomiya,
Yukihiko Satou, Keisuke Sueki and
Manabu Fukumoto

Chapter 17   **Immortalization of Radiation-Exposed Tissues**   **206**
Tomokazu Fukuda

Chapter 18   **Water with Radioactive Cs: Impacts on Mice**   212
Hiroo Nakajima

Part VI   **Fukushima Accident: A Review**   227

Chapter 19   **Biological Impacts of FNPP Accident: A Study of Pale Grass Blue Butterfly**   229
Joji M. Otaki

Chapter 20   **Fukushima and Chernobyl: A Comparative Study**   238
Tetsuji Imanaka

Permissions

Index

PREFACE

The main aim of this book is to educate learners and enhance their research focus by presenting diverse topics covering this vast field. This is an advanced book which compiles significant studies by distinguished experts in the area of analysis. This book addresses successive solutions to the challenges arising in the area of application, along with it; the book provides scope for future developments.

Radioecology refers to the study of ecology concerned with the presence of radioactivity in Earth's ecosystems. Radioecology research includes experimental laboratory and field procedures, field sampling and the construction of environmentally predictive simulation models in an effort to understand how radioactive material migrates through the environment. The practice combines techniques from different disciplines such as mathematics, physics, ecology, chemistry and biology. It is applied for the protection of different organisms from radiation. DNA is the major target in all the living organisms for the initiation of biological effects due to radiation. There exist wide similarities and differences in radiation responses in various organisms. Radioecological studies provide the information required for risk assessment and dose estimation in the context of radioactive pollution and its effects on human health and the environment. The topics included in this book on radioecology are of utmost significance and bound to provide incredible insights to readers. It provides significant information of this discipline to help develop a good understanding of radiation dynamics and its impact. The book is appropriate for students seeking detailed information in this area of study as well as for experts.

It was a great honour to edit this book, though there were challenges, as it involved a lot of communication and networking between me and the editorial team. However, the end result was this all-inclusive book covering diverse themes in the field.

Finally, it is important to acknowledge the efforts of the contributors for their excellent chapters, through which a wide variety of issues have been addressed. I would also like to thank my colleagues for their valuable feedback during the making of this book.

Marty Oates

1

Understanding the Consequences of Radiation for Ecosystem and Human Health

Manabu Fukumoto

Abstract The adverse effect of radiation on human health, especially cancer induction, is a major concern, especially after the Fukushima Daiichi Nuclear Power Plant (FNPP) accident. We have learned the consequences of radiation for human health through radiological tragedies and nuclear disasters. Archival materials on Thorotrast patients have enabled us to perform molecular pathological analysis, in order to elucidate the carcinogenic mechanism of internal radiation exposure. Radiation-induced cancer is not merely attributed to resulting genetic mutations, but is in fact a complex consequence of the biological response to radiation and ingested radionuclides. Therefore, the FNPP accident prompted us to launch "A comprehensive dose evaluation project on animals affected by the FNPP accident" to establish an archive system composed of samples and data from animals around FNPP. Using those archived samples, we have been able to report some achieved results. The final goal of this archive system is to enable research that will contribute to the common understanding of the radioprotection of the ecosystem as well as humans.

In reality, however, it is becoming difficult to continue this project, due to reduced research spending at academic institutions and the weathering memory for the accident of people.

Keywords Thorotrast · Radiation exposure · Fukushima Daiichi Nuclear Power Plant accident · Evacuation zone · Livestock · Japanese macaque

1.1 Introduction

Exposure to radiation above a certain level obviously has adverse effects on human health. Radiation is not detectable by our five senses, and half of people exposed to 4 Gy of photons would die within 60 days if the whole body were exposed, which

M. Fukumoto (✉)
Institute of Development, Aging and Cancer, Tohoku University, Sendai, Japan

School of Medicine, Tokyo Medical University, Tokyo, Japan
e-mail: manabu.fukumoto.a8@tohoku.ac.jp

Questions frequently asked
- "No immediate effects" ; Does it mean secure?
  How about the risk of cancer? No impact on offspring?
- Are low levels of radiation good for health?
- Is artificial radiation bad, but Radon hot spring good for health?
- Is internal exposure more dangerous than external one?

Facts with certainty
- Total body exposure to 4 Gy: LD50/60 (half die within 60 days by the small energy equivalent to increasing 0.001°C of body temperature)
  Cancer risk increases if survived acute exposure
- Cancer risk is higher if exposed earlier.
- Human effects of exposure to < 0.1Gy are unknown.
- Radiation specific pathological changes are unknown.

Fig. 1.1 Radiation effects on humans; what people want to know is hard to know and what we know is limited

is only enough energy to raise the body temperature by 0.001 °C [1]. There is, therefore, a general belief that ionizing radiation is dangerous at any dose. The world's most reliable data about the effect of radiation on human health is the epidemiological survey on atomic-bomb survivors of Hiroshima and Nagasaki (Hibakusha) [2]. After radiation exposure, symptoms occurring within months are called acute (or early) effects, and those that develop after years to decades of a symptomless incubation period are called late (or delayed) effects. A life span study of Hibakusha (LSS) revealed that the incidence of cancer (relative cancer risk), which is a late effect, was proportional to dose. The Fukushima Daiichi Nuclear Power Plant (FNPP) accident has not caused acute effects on people in general; however, it is difficult to predict whether late effects will appear in the future. The most concerning of these late effects is cancer (Fig. 1.1).

1.2   External and Internal Radiation

The data from Hibakusha relate to an instantaneous external exposure to high-dose-rate radiation, but cannot speak to long-term internal exposure at low-dose-rate (LDR). Thorotrast is an angiographic contrast medium composed of a colloidal solution of thorium dioxide, which is a natural α-particle emitter. It was administered to wounded soldiers during World War II, and more than half of Thorotrast accumulated in the liver. Liver cancers have been evoked by this internal exposure decades after the administration. Thorotrast-induced liver cancers include intrahepatic cholangiocellular carcinoma (arising from epithelial cells of the bile duct), angiosarcoma (from vascular endothelial cells), and hepatocellular carcinoma (from liver parenchymal cells) and occur at a frequency of about 3:2:1, respectively. More than 80% of non-Thorotrast liver cancer is hepatocellular carcinoma, while the rate of angiosarcoma is negligible. These differences show how Thorotrast-induced liver cancer is crucial to understand the oncogenic mechanisms of persistent LDR internal radiation in humans [3]. Supported by the War Victims' Relief Bureau, former Ministry of Health and Welfare of Japan, approximately 400 Thorotrast patients underwent the postmortem pathological examination. Paraffin-embedded blocks, clinical course, thorium concentration and other information were comprehensively gathered. This systematic archive is valuable for elucidating the molecular mechanisms of human cancers induced by internal radiation exposure, particularly as it is composed of the world's largest number of related cases with the highest quality of radiological and pathological information [4]. The profile of cancer induction is quite different between Hibakusha and Thorotrast patients, indicating that the mechanisms underlying the carcinogenic effect of radiation vary between external and internal exposure and, therefore, have to be analyzed differently [3]. LSS revealed that the likelihood of cancer is different depending on the originating organ [5]. However, pathological characteristics specific to radiation carcinogenesis have not been found. In disasters, such as the aftermath of nuclear weapons and nuclear power plant accidents, released radioactive materials cause both external and internal radiation exposure. Once radioactive materials are ingested, they are not evenly distributed throughout the body, but become enriched in specific organs dependent on their chemical and physical properties, and the target organs are heavily exposed. Even in an organ, the distribution of radionuclides and irradiation at the microscopic level is not homogeneous [6]. In addition, deposited radionuclides form a much more complicated exposure profile than in external exposure, due to physical radioactive decay, biological excretion of radioactive materials [7], and dynamic remodeling of the organ [8].

1.3   The Biological Effect of Radiation

From the classical radiobiological standpoint, it is necessary to clarify the existence of a dose-effect relationship, in order to prove that a certain biological phenomenon (effect) is attributed to radiation. Radiation exerts no effect unless its energy is absorbed in the cell. Therefore, it is impossible to discuss the biological effect of radiation unless specifying both the quality and dose of radiation. A variety of biological effects of radiation are principally explained as being caused by a particular series of events. A cell is injured directly by radiation or indirectly by reactive oxygen species produced by radiation. Among the injured cellular components, DNA double-strand breaks (DSBs) are considered to have the most serious effect on the cell. DSBs undergo the repair process and can be divided into the following three patterns: (1) cell death due to irreparable breaks induced by high intensity/ dose radiation, (2) genetic alterations due to incomplete repair, or (3) recovery to a normal cell with intact, unmutated DNA (Fig. 1.2). Cell death results in deterministic acute effects, while genetic alterations result in stochastic effects such as cancer induction and transgenerational (or hereditary) effects. Among stochastic effects,

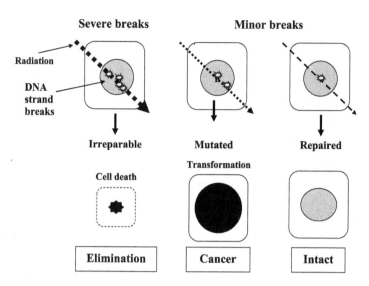

Fig. 1.2 Traditional but too much simple thinking of cellular effects of radiation

cancer induction is people's major concern and is thought to be a consequence of accumulated genetic mutations and the clonal expansion of a cell toward a cancerous state. Cancer development takes a long period of time and is therefore a late effect of radiation. Here arises a big question: what conditions does radiation work toward either the direction of cell death or cell immortality (carcinogenesis), two extreme events?

There is still much that is unknown about the effects of radiation. Radiation damage to DNA is relatively well studied, but damage to other cell components and epigenetic changes also need to be better understood [9]. Since cells have coping and restoration mechanisms, it is understandable that long-term exposure to LDR radiation is less harmful than acute exposure to high-dose-rate radiation when exposed to the same cumulative dose. However, this may not be true to all situations [10]. The effect of radiation is the result of complex biological reactions including the non-targeting effect [11] and all effects, from cellular changes beginning at exposure to overt changes, must be considered. These concepts indicate that the abovementioned three-way model (Fig. 1.2) is too simple to explain the multifaceted nature of radiation damage. Several problems are to be considered when cancer biology is analyzed (Figs. 1.3 and 1.4).

Both irradiated and bystander cells can be the cell of origin for radiation-induced cancer. There is no clear consensus as to whether internal and external exposures exert the same biological effect if the dose is equal. Even the organ dose of plutonium after the nuclear bombing of Nagasaki Hibakusha was evaluated to be far lower than external exposure; however, the impact to the individual cell nucleus by a single α-particle is not negligible [12]. We should, therefore, remember that there is great uncertainty in dose evaluation itself, especially in internal dose assessment.

1. Internal exposure & External exposure
(different dose distribution)

2. Leukemia & Solid cancers
(molecular mechanisms of carcinogenesis)

3. Cell death & Cancer formation
(Cell immortalization)
(Acute effects & Late effects)

4. Cells irradiated & Cell of origin for cancer
(tissue stem cells, bystander effect, adaptive response)

5. Dose/dose-rate needed for cancer induction
(dose accumulation / dose evaluation)

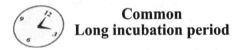

Common
Long incubation period

Fig. 1.3 In order to elucidate the mechanisms of radiation carcinogenesis, various conditions should be considered. However, it is common that radiation carcinogenesis takes a long time, no matter what kind of exposure

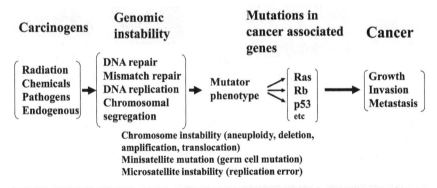

Recent paradigms → Cancer could be also derived from an unirradiated cell.

1. Cancer stem cells
2. Non-targeted effects (bystander effects, adaptive responses)
3. Epigenetic changes (methylation, acetylation, RNA interferences)

Fig. 1.4 Carcinogenic pathway. Molecular changes in radiation carcinogenesis are common among different carcinogens. Molecular changes proved in thorotrast-induced liver tumors are written in red ink. The paradigm shift in radiation biology and cancer biology shows the possibility that a cell not directly irradiated can be a cell of origin of cancer

In addition, we experience problems with the units related to radiation. Radioactivity is measured by how many radioactive decays occur per second, for which the Becquerel (Bq) unit is used. The unit of absorbed dose is gray (Gy, J/kg). Mass communications generally use sievert (Sv) instead of Gy to express radiation dose. Sv takes all the conditions into consideration for evaluating the effect on the human body, including modifiers such as radiation quality, exposure period, the differences between exposed organs, and internal/external exposure. Using conversion factors, physically measured Bq is converted to the biological unit Sv. The totally different concepts of equivalent dose, effective dose, and committed dose for a person's lifetime are also expressed by the same unit, Sv [13]. However, it is ambiguous whether the conversion factors from other units to Sv have the appropriate numerical value. Sv was originally elaborated as a measure of the health effect of low levels of ionizing radiation on the human body but is used for the even deterministic effect which is a discontinuous event from the stochastic effect. Many people, including researchers, misunderstand that Sv exists as a physical unit for measuring absorbed dose since the unit was proposed. Equivalent dose expressed by Sv may be measurable by a monitor but is modified to account for the effectiveness of the type of radiation based on the absorbed dose. It is emphasized again that the final aim of Sv is not for general dose estimation, but for the protection of human health.

1.4   The FNPP Accident and Our Project

Following the Great East Japan Earthquake on March 11, 2011, the FNPP accident released a large amount of radioactive materials into the environment. When farmers within a 20-km radius from FNPP were ordered to evacuate shortly after the beginning of the disaster, livestock were released due to unprecedented confusion and compassion for the animals. After the area was defined as the evacuation zone on April 22, 2011, livestock formed herds and roamed empty streets (Fig. 1.5). There was concern that contaminated meat would come on the market. Therefore, on May 12, 2011, the prime minister ordered Fukushima Prefecture to euthanize livestock within the evacuation zone. I came to the idea that the livestock intended for death should, instead, be used to study radiation protection for humans. Since experiments of this type are impossible, effects of radiation exposure on humans and the ecosystem can be learned only through radiation and nuclear disasters. It is undeniable that studies of radiation biology and protection have progressed by the accumulation of data from major disasters, such as Hibakusha and the Chernobyl NPP (CNPP) accident. We should not, of course, cause nuclear accidents, but we need to learn as much as possible from the FNPP accident. In other words, in the face of the unprecedented pollution over such a wide area, investigating the various impacts of the FNPP accident and conveying these analytical and strategic findings to posterity is a responsibility that has been imposed on the scientific community in Japan. Therefore, we launched "A com-

Fig. 1.5 The ex-evacuation zone set on April 22, 2011. (**a**) Gate set at Kawauchi Village (September 28, 2011); livestock formed herds and roamed empty streets. (**b**) Cattle became wild in front of a house damaged by the tsunami (Tomioka Town January 25, 2012), (**c**) wild boars (Tomioka Town, February 28, 2012), (**d**) unleashed pigs (Tomioka Town, January 25, 2012). (The photos are provided by courtesy of Prof. H. Yamashiro)

prehensive dose evaluation project concerning animals affected by the Fukushima Daiichi Nuclear Power Plant accident" [14]. Following tough negotiations with administrative agencies, nearly half a year after the accident, we were finally allowed to enter the evacuation zone and perform sampling on August 29, 2011. The sampling of livestock continued until the end of March 2013, at which time all the livestock in the zone were euthanized. After that, we have concentrated on research using archives and have shifted to the sampling of wild Japanese macaques [15].

1.5   Results of the Project to Date

Several achievements from the project have already been published and are briefly described below.

1. In all of the organs of any of affected animals, the deposition of radioactive cesium (Cs) was observed and was found to be the highest in skeletal muscle. A

linear correlation between the radioactive Cs concentration in the peripheral blood and in each organ was also determined. The levels of radioactive Cs in the organs of fetuses and infants were higher than in the corresponding maternal organs. The organ-specific deposition of radionuclides with relatively short half-lives was detected, including silver-110 m (half-life: 249.8 days) in the liver and tellurium-129 m (33.6 days) in the kidneys [16].

2. Plasma levels of malondialdehyde and superoxide dismutase activity in affected cattle were positively correlated, and glutathione peroxidase activity was inversely correlated with the internal dose rate of radioactive Cs, suggesting that chronic exposure to LDR radiation induces mild oxidative stress in the affected cattle [17].

3. DNA double-strand breaks in peripheral blood lymphocytes of affected cattle were determined by immunocytochemical staining for γ-H2AX. While the extent of DNA damage appeared to be independent of the distance from FNPP and the estimated radiation dose from radioactive Cs, we observed age-dependent accumulation of DNA damage. The levels of DNA damage decreased slightly over the 472-day sample collection period. When analyzing long-term exposure effects of LDR radiation, it is necessary to consider the effect of adaptive response and aging of individuals [18].

4. Testis and bone marrow are highly radiosensitive. However, no morphological changes were detected in the testis of cattle, boar, or inobuta (wild boar and domestic pig hybrid) stayed in the ex-evacuation zone for about 1 year after the accident [19, 20]. Spermatogenesis in large Japanese field mice was found to be enhanced [21], and internal dose-rate-dependent myelosuppression was present in the bone marrow of adult wild macaques without obvious health effects [15]. It remains to be elucidated whether these phenomena, attributed to chronic exposure to LDR radiation, will benefit or adversely affect animals.

5. The expression of genes related to immunity was altered in the small intestine of affected swine and inobuta. Chronic LDR radiation may evoke a persistent slight inflammatory status [22].

6. Strontium-90 (^{90}Sr) was detected in the teeth and bones of affected cattle and could provide useful information about internal exposure. The fluctuation in the teeth is suggested to reflect the contamination levels of environmental ^{90}Sr [23].

7. Using 2-D differential gel electrophoresis and time-of-flight (TOF) mass spectrometry, a proteome analysis of rice leaves revealed that the majority of the differentially expressed proteins after low-level gamma radiation were within the general (non-energy) metabolism and stress response categories [24].

As a result of this project, the biological impact of the FNPP accident on animals has been shown to be subtle; the change was most strongly correlated with internal dose-rate among the four combinations of indicators, external/internal and dose/dose-rate. Overall, obvious adverse effects have, so far, been undetectable, which is thought to be due to the adaptability of animals. However, it remains to be elucidated whether long-term exposure to LDR radiation will impair the ability of animals to compensate in the future. Every summer since 2013, we have held a meeting, gathering researchers from all over Japan to seriously study problems in relation to the FNPP accident. The meeting is open to the public, as well, and we present up-to-date results, host frank scientific discussion about the problem, and share information [25].

1.6   Current Issues and Future Prospects

Radiation causes a significant increase in the incidence of thyroid cancer [26]. Increased incidence of thyroid carcinoma due to iodine-131 (^{131}I, half-life: 8.02 days) from the CNPP accident, which scattered about ten times as much radioactive materials as the FNPP accident, was not noticed until about 5 years after the accident [27]. If the occurrence frequency of the radiation effect is dependent on the total dose, even 50-year follow-up observation is necessary for the FNPP accident. From our dose evaluation, the initial exposure by short half-life radionuclides such as ^{131}I was not negligible (data not shown). The ratio of ^{131}I concentration to ^{129}I (half-life: 1.57×10^7 years) concentration released by the FNPP accident is known. Using accelerator mass spectrometry, ^{129}I is measurable, and a retrospective reconstruction of ^{131}I levels from the levels of accident-derived ^{129}I was successful [28]. We also believe that the dose attributed to ^{131}I can be evaluated from ^{129}I concentration in the thyroid gland. The lens is relatively radio-sensitive, and radiogenic cataracts have been documented as a major ocular complication, one of the deterministic but late effects. Their threshold dose and the incubation period are now of major concern in relation to the radiation effect [29]. Careful observation must be maintained, in order not to overlook even small changes in animals. If any ecological change is found, it will be a warning to consider the potential impact on humans. Various changes in the ecosystem after the FNPP accident have been reported. However, it is often unknown whether they are really due to radiation, since these changes have not necessarily been reported in association with dose or dose rate. As mentioned elsewhere, efforts are needed to create links between laboratory-controlled experiments and in situ field analysis in cooperation with radiation biologists and ecologists for the establishment of trusted radioecology data [30].

Currently, we are focusing on sampling materials from wild Japanese macaques as the main subject of our analysis. As seen in macaques inhabiting the area affected by the FNPP accident, wildlife contamination by radioactive Cs is continuing even now, 8 years after the accident (Fig. 1.6), which demonstrates that animals are still subject to the so-called high-dose-rate environment. To understand the human response to long-term exposure to very LDR radiation, Japanese macaques are the most suitable wild animal, since they do not recognize or fear radiation and do not

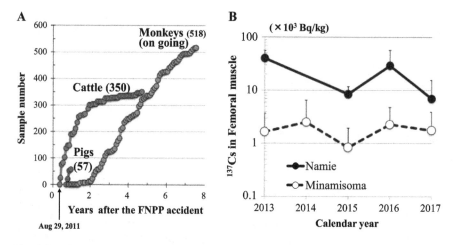

Fig. 1.6 Wild Japanese macaques (monkeys) are the most suitable to know radiation effects on humans. (a) Cumulative numbers of animals sampled in this project (as of October 25, 2018). (b) Wildlife contamination by radioactive cesium is continuing even now without signs of falling down in the affected area

smoke or drink, which is the most confounding factor in analyzing the effects of low-dose (LD) radiation on humans. The life span of Japanese macaques is approximately 20 years, and many individuals born after the FNPP accident have been observed. Therefore, they provide extremely crucial data for understanding the effect of chronic, very LDR radiation on humans including transgenerational effects and should be vigilantly monitored over time. We started evaluating the cumulative dose of individual macaques by electron spin resonance (ESR) analysis of their teeth. Despite the fact that macaques are the closest species to humans, they are not among the reference animals chosen by ICRP [31], because the areas where nuclear disasters have occurred in the past have not been inhabited by macaques. Therefore, the administrative agencies of Japan judge that the results obtained in wild Japanese macaques cannot be compared with other related study results, which disappointingly means that they officially hold no significance as the subject of comparative analysis.

At the time they were taken, it was unknown whether the archived pathological specimens of Thorotrast patients would be useful for the future. However, great strides in technology have allowed us to perform genetic analysis using old paraffin-embedded blocks. With PCR, we performed genetic analysis on the archived paraffin blocks of Thorotrast-induced liver tumors decades after autopsy. We showed that gene mutations are not caused by the direct action of radiation, but as a result of the biological reaction to radiation [3]. As mentioned above, we are currently achieving successful results with the archived samples of affected animals. Maintaining the archive in a form that any researcher can use makes it possible to detect small effects attributed to long-term exposure to LDR radiation that are not currently known. For example, results acquired by comprehensive gene analysis using next-generation DNA sequencing or epigenetic analysis will extend the knowledge of how radiation influences human health and is scientifically necessary for further radioprotection efforts. The biological impact of radiation is influenced by various factors such as dose, dose-rate, and quality of radiation and species, age and gender (if any), and so on of organisms. Therefore it is difficult to scientifically describe the influence of LD/LDR radiation. In the future, new knowledge will be enhanced by analyzing research data accumulated with deep learning of artificial intelligence. For those purposes, accumulation of detailed data as accurate as possible is desired.

Our project after the FNPP accident has two key trajectories concerning the effect of LDR radiation. One is the fear of what will ultimately happen to the natural environment due to the FNPP accident. The other is the need for field studies to clarify the influence of long-term LDR radiation on the natural environment, which is now in progress. Regardless of which goal we adopt, it is clear that we need to think about the protection of the environment and humans after radiological accidents [32].

Acknowledgments I sincerely appreciate all the people and researchers who have been involved in the project. This project became executable with the support of the Emergency Budget for the Reconstruction of Northeastern Japan, MEXT, Japan, Grants-in-Aids for scientific research from JSPS (Kakenhi 26253022, 15H01850) Discretionary Expense of the President of Tohoku University, and the Program for the Promotion of Basic and Applied Research for Innovations in Bio-oriented Industry.

References

1. Sakata R, Grant EJ, Ozasa K (2012) Long-term follow-up of atomic bomb survivors. Maturitas 72(2):99–103
2. Hall EJ, Giaccia AJ (eds) (2012) Radiobiology for the radiologist, 7th edn. Lippincott Williams & Wilkins, Wolters Kluwer, Philadelphia
3. Fukumoto M (2014) Radiation pathology: from thorotrast to the future beyond radioresistance. Pathol Int 64(6):251–262
4. Database on thorotrast patients in Japan. http://www2.idac.tohoku.ac.jp/misc/thorotrast/index%20english.html
5. Radiation Effects Research Foundation (RERF). Life Span Study (LSS) Report Series. https://www.rerf.or.jp/en/library/list-e/scientific_pub/lss/
6. Yamamoto Y, Usuda N, Oghiso Y et al (2010) The uneven irradiation of a target cell and its dynamic movement can mathematically explain incubation period for the induction of cancer by internally deposited radionuclides. Health Phys 99(3):388–393
7. Yamamoto Y, Chikawa J, Uegaki Y et al (2010) Histological type of thorotrast-induced liver tumors associated with the translocation of deposited radionuclides. Cancer Sci 101(2):336–340

8. Yamamoto Y, Usuda N, Takatsuji T et al (2009) Long incubation period for the induction of cancer by thorotrast is attributed to the uneven irradiation of liver cells at the microscopic level. Radiat Res 171(4):494–503

9. Zhou H, Hong M, Chai Y et al (2009) Consequences of cytoplasmic irradiation: studies from microbeam. J Radiat Res 50(Suppl A):A59–A65

10. Tomita M, Maeda M (2015) Mechanisms and biological importance of photon-induced bystander responses: do they have an impact on low-dose radiation responses. J Radiat Res 56(2):205–219

11. Mothersill C, Seymour C (2018) Old data-new concepts: integrating "Indirect Effects" into radiation protection. Health Phys 115(1):170–178

12. Shichijo K, Takatsuji T, Fukumoto M et al (2018) Autoradiographic analysis of internal plutonium radiation exposure in Nagasaki atomic bomb victims. Heliyon 4:e00666

13. ICRP (2007) The 2007 Recommendations of the International commission on radiological protection. ICRP publication 103. Ann ICRP 37(2–4):1–332

14. Takahashi S, Inoue K, Suzuki M et al (2015) A comprehensive dose evaluation project concerning animals affected by the Fukushima Daiichi Nuclear Power Plant accident: its set-up and progress. J Radiat Res 56(Suppl 1):i36–i41

15. Urushihara Y, Suzuki T, Shimizu Y et al (2018) Haematological analysis of Japanese macaques (*Macaca fuscata*) in the area affected by the Fukushima Daiichi Nuclear Power Plant accident. Sci Rep 8:16748

16. Fukuda T, Kino Y, Abe Y et al (2013) Distribution of artificial radionuclides in abandoned cattle in the evacuation zone of the Fukushima Daiichi nuclear power plant. PLoS One 8(1):e54312

17. Urushihara Y, Kawasumi K, Endo S et al (2016) Analysis of plasma protein concentrations and enzyme activities in cattle within the ex-evacuation zone of the Fukushima Daiichi Nuclear Plant Accident. PLoS One 11(5):e0155069

18. Nakamura AJ, Suzuki M, Redon CE (2017) The causal relationship between DNA damage induction in bovine lymphocytes and the Fukushima Nuclear Power Plant accident. Radiat Res 187(5):630–636

19. Yamashiro H, Abe Y, Fukuda T et al (2013) Effects of radioactive caesium on bull testes after the Fukushima nuclear plant accident. Sci Rep 3:2850

20. Yamashiro H, Abe Y, Hayashi G et al (2015) Electron probe X-ray microanalysis of boar and inobuta testes after the Fukushima accident. J Radiat Res 56(Suppl 1):i42–i47

21. Takino S, Yamashiro H, Sugano Y et al (2017) Analysis of the effect of chronic and low-dose radiation exposure on spermatogenic cells of male large Japanese field mice (*Apodemus speciosus*) after the Fukushima Daiichi Nuclear Power Plant accident. Radiat Res 187:161–168

22. Morimoto M, Kato A, Kobayashi J et al (2017) Gene expression analyses of the small intestine of pigs in the ex-evacuation zone of the Fukushima Daiichi Nuclear Power Plant. BMC Vet Res 13(1):337

23. Koarai K, Kino Y, Takahashi A et al (2016) (90)Sr in teeth of cattle abandoned in evacuation zone: record of pollution from the Fukushima-Daiichi Nuclear Power Plant accident. Sci Rep 6:24077

24. Hayashi G, Moro CF, Rohila JS et al (2015) 2D-DIGE-based proteome expression changes in leaves of rice seedlings exposed to low-level gamma radiation at Iitate village, Fukushima. Plant Signal Behav 10(12):e1103406

25. Fukumoto M, Imanaka T (2015) The first critical workshop on the effect of the Fukushima Daiichi Nuclear Power Plant accident on the ecosystem and on humans. J Radiat Res 56(Suppl 1):i1

26. Albi E, Cataldi S, Lazzarini A et al (2017) Radiation and thyroid cancer. Int J Mol Sci 18(5). pii:E911

27. Demidchik YE, Saenko VA, Yamashita S (2007) Childhood thyroid cancer in Belarus, Russia, and Ukraine after Chernobyl and at present. Arq Bras Endocrinol Metabol 51(5):748–762

28. Fujiwara H (2016) Observation of radioactive iodine (131I, 129I) in cropland soil after the Fukushima nuclear accident. Sci Total Environ 566–567:1432–1439

29. Chodick G, Bekiroglu N, Hauptmann M et al (2008) Risk of cataract after exposure to low doses of ionizing radiation: a 20-year prospective cohort study among US radiologic technologists. Am J Epidemiol 168(6):620–663
30. Bréchignac F, Oughton D, Mays C et al (2016) Addressing ecological effects of radiation on populations and ecosystems to improve protection of the environment against radiation: agreed statements from a Consensus Symposium. J Environ Radioact 158–159:21–29
31. ICRP (2008) Environmental protection – the concept and use of reference animals and plants. ICRP publication 108. Ann ICRP 38(4–6)
32. Oughton DH (2016) Ethical foundations of environmental radiological protection. Ann ICRP 45(1 Suppl):345–357

Part I
The Impact of Fukushima
Nuclear Power Plant Accident on
the Environment

2

Fukushima Accident: Radiation Exposure to Mice

Manabu Onuma, Daiji Endoh, Hiroko Ishiniwa, and Masanori Tamaoki

Abstract Following the Fukushima Daiichi Nuclear Power Plant (FNPP) accident, dose rate for Muridae species in forests of Iitate Village, Fukushima, was estimated as 3.9 mGy/day over the first 30 days. According to the derived consideration reference levels (DCRLs) determined by the International Commission on Radiological Protection (ICRP), this dose level could be affecting reproduction of these rodents. However, information on dose rate for forest rodents after 2012 is limited. Therefore, the dose rate of forest rodents was calculated for large Japanese field mice (*Apodemus speciosus*) captured in the "difficult-to-return zone" in the Fukushima Prefecture from 2012 to 2016. External dose rate was calculated based on the ambient dose equivalent rate of gamma-radiation at the ground level of the trapping site. Internal dose rate was simulated using the EGS5 program based on cesium (Cs)-137 concentrations in the captured mice. Combining the external and internal doses, the total daily dose rate for the mice within the zone was estimated to be 0.201–0.547 mGy/day. In addition, the ratio of external dose rate to total dose rate was estimated to be 61.2–95.4%. Thus, it is concluded that the present radiation exposure of the field mice distributed in the trapping site did not affect their reproduction. However, it must be noticed that total dose rate exceeding 0.1 mGy/day, which offers very low

M. Onuma (✉)
Ecological Risk Assessment and Control Section, Center for Environmental Biology and Ecosystem Studies, National Institute for Environmental Studies, Tsukuba, Japan
e-mail: monuma@nies.go.jp

D. Endoh
Department of Radiation Biology, School of Veterinary Medicine, Rakuno Gakuen University, Ebetsu, Japan

H. Ishiniwa
Institute of Environmental Radioactivity, Fukushima University, Fukushima City, Fukushima, Japan

M. Tamaoki
Fukushima Branch, National Institute for Environmental Studies, Miharu Town, Tamura County, Fukushima, Japan

probability of the occurrence of certain effects according to the DCRLs determined by ICRP, is still present in most of the zone (September 2018, Nuclear Regulation Authority, Japan). Thus, various indexes should be applied to evaluate the exposure effects on the field mice in this zone.

Keywords Large Japanese field mouse · *Apodemus speciosus* · Does rate · EGS5 · Cs-137

2.1 Introduction

The Great East Japan Earthquake occurred on March 11, 2011, 14:46 JST, with its epicenter at N38.1, E142.9 (130 km ESE off Oshika Peninsula) and a depth of 24 km [1]. As an aftermath of the earthquake, the coastal area of the main island, Honshu, was affected by tsunami, which resulted in loss of power supply at Fukushima Daiichi Nuclear Power Plant (FNPP). After that, reactors were damaged and meltdown occurred. Eventually, substantial amounts of radioactive materials were released into the surrounding environment. The total amount of radioactive materials released into the environment was estimated by several organizations such as the Nuclear and Industrial Safety Agency (NISA), the Nuclear Safety Commission (NSC) and Tokyo Electric Power Co. (TEPCO) [2–6]. The estimated values of iodine-131 (^{131}I) ranged between 1.2 and 5.0×10^{-17} Bq. The value of cesium-137 (^{137}Cs) was calculated to be between 8.2 and 15.0×10^{-15} Bq. The released radioactive materials formed radioactive plume, some parts of which moved into the inland, mainly Fukushima Prefecture. Radioactive materials were dissolved in rainwater contaminated soil. For example, ^{131}I (T1/2 = 8 days, 71–28 kBq/kg), ^{137}Cs (T1/2 = 30.0 years, 10–43 kBq/kg), and ^{134}Cs (T1/2 = 2.06 years, 8.5–42 kBq/kg) were detected in 5 cm surface of soil collected from five locations of Iitate Village located 25–45 km northwest of FNNP, in March 2011 [7]. During the early phase following the accident, radioactive I was the main contaminant of the ground surface. Thereafter, ^{137}Cs became the main contaminant because of a longer half-life. ^{137}Cs decays by emitting β-particles and both Cs-137 and its metastable nuclear isomer barium-137m (^{137m}Ba) emits γ-rays [8]. In addition, ^{137}Cs binds to clay firmly and the clay-bound ^{137}Cs has low mobility [9, 10]. Therefore, radiation from ^{137}Cs could affect biological organisms due to long-term exposure to the contaminated environment.

The International Commission on Radiological Protection (ICRP) has recommended some organisms called reference animals and plants for environmental protection with regard to radiation [11]. Deer (family Cervidae), rat (family Muridae), duck (family Anatidae), frog (family Ranidae), bee (family Apidea), earthworm (family Lumbricidae), pine tree (family Pinacea) and wild grass (family Poaceae) are designated as reference animals and plants for the terrestrial environment. Rat (family Muridae) offers most information on radiation exposure than any other reference animals because laboratory rats and laboratory mice have been used for laboratory experiments in various kinds of researches, including radiation biology. Following the FNPP accident, the dose rate for rodents in the forest in Fukushima Prefecture was reported as 3.9 mGy/day over the first 30 days [12]. According to the derived consideration reference levels (DCRLs) determined by ICRP, this dose level could be affecting reproduction, due to reduced fertility in male and female of Muridae species. However, a major portion of the radiation source changed from ^{131}I to mainly ^{137}Cs because of the decay of radioactivity, thus, it became necessary to evaluate radionuclides influence on the reproduction of Muridae species in Fukushima.

2.2   Materials and Methods

2.2.1   Animals

The Large Japanese field mouse (*Apodemus speciosus*) (Fig. 2.1), hereafter called as the field mouse, was selected as the target species. This species belongs to family Muridae and is a common rodent endemic to Japan. Thus, the species fulfills the criteria of reference animlas specified by ICRP and mice exposed and nonexposed to ^{137}Cs are easily obtained owing to its large distribution area in Japan.

The general information of the field mouse is as described below [13]. The ranges of their external measurements are as follows: head and body lengths, 80–140 mm; tail length, 70–130 mm; hindfoot length (not including claw), 22–28 mm; and body weight, 20–60 g. The species inhibits in Hokkaido, Honshu, Kyushu and islands larger than 10 km^2. The habitats of the species are forests, plantations, riverside fields with dense grass, paddy fields and cultivated fields. Their diet consists of root and stems of herbaceous plants, nuts, berries and small insects. The breeding season varies geographically, from April to September in Sapporo City (Hokkaido), from March to April and September to November in Kyoto Prefecture (Honshu), and from October to March in Nagasaki Prefecture (Kyushu). The gestation period is 19–26 days under laboratory condition. The average litter size is 4.0–6.2 in the wild. The reported maximum life span in the wild is estimated to be 15 months in Hokkaido and 26 months in Honshu based on live trapping.

Fig. 2.1 The large
Japanese field mouse
(*Apodemus speciosus*)

Three chromosome types are observed in these species, 2n = 46 type is distributed in Western Japan, and 2n = 48 is distributed in Eastern Japan. The hybrid type 2n = 47 occurs on boundary of both types and is the so-called Kurobe–Hamamatsu Line.

2.2.2   External Dose Rate

External dose rate was calculated from daily external dose rate based on the ambient dose equivalent rate of gamma-radiation at ground level of the trapping site.

2.2.2.1   Characteristic of Trapping Site

The trapping site (latitude, 37° 36′ 02″; longitude, 140° 45′ 07″ E; altitude, 578 m) was located northwest of FNPP in Fukushima Prefecture. The vegetation at the site is mixed coniferous and broadleaf forest. The site was within the "difficult-to-return zone" (hereafter the DR zone). The ambient dose equivalent rate was measured at randomly chosen spots of the trapping site. The number of the measurement spots was four in 2012 and five from 2013 to 2016.

The ambient dose rate of gamma-radiation at ground level were determined using a portable environmental gamma survey meter (NHE20CY3-131By-S, Fuji Electric Co., Ltd., Tokyo, Japan) which is in full compliance with JIS Z 4333 (2006) (the conversion factor of ambient dose equivalent, $H*(10)$ for 662 keV γ-photons from ^{137}Cs is 1.2 (Sv/Gy)).

2.2.2.2   Calculation of External Dose Rate

Being the radiation converting factor of γ-ray 1.0, the daily external dose rate based on the ambient dose equivalent rate was calculated using the following formula:

Daily external dose rate (mGy/day) = (Ambient dose equivalent rate of trapping site (μSv/h)/1.2) x 24 hrs/1 (radiation weighting factor of γ-ray)/1000

2.2.3   Internal Dose Rate

The daily internal dose rate based on ^{137}Cs concentration of the field mouse body was calculated using Monte Carlo electron-photon transport code EGS5 [14].

2.2.3.1   Trapping of the Field Mice

Trapping of the field mice was conducted by Sherman-type live traps baited with sunflower seeds. The traps were set from August to November 2012, from July to October 2013, from June to August 2014, in August 2015, and from August to September 2016. The traps were observed the following day and the trapped field mice were euthanized using CO_2 asphyxiation.

The field mice were handled in accordance with the guidelines for studying wild mammals of the Mammal Society of Japan [15] and the rules of the National Institute for Environmental Studies for analysis and experimentation with environmental samples contaminated with radioactive materials. The permission for trapping the field mice was obtained from the Ministry of the Environment, Ministry of Agriculture, Forestry and Fisheries and Fukushima Prefectural Office.

2.2.3.2   Measurement of ^{137}Cs Concentrations in the Field Mice

The age of the euthanized field mice was determined based on tooth wearing [16]. The field mice having tooth wear stage 4 or above (>5 month in age; >23.3 g in female; >27.0 g in male) were used to measure ^{137}Cs concentrations. The head and internal organs were removed from the body and the processed bodies were stored at −20 °C until the next processing. The processed body parts consisting of bones, muscles and fur skin, were minced individually using a food processer. These minced body parts were mixed well and then transferred to a polystyrene container (U8 container). High-purity germanium (HpGe) detectors (GMX45P4-76, ORTEC, TN, or GCW7023, CAMBERRA industries Inc., TN) calibrated by a standard source (MX033U8PP, the Japan Radioisotope Association) were used to measure ^{137}Cs activity in each minced body. Gamma Studio (SEIKO EG&G CO., LTD., Tokyo, Japan) and Spectrum Explorer (CAMBERRA) software were used to analyze the γ-ray spectra. Cs-137 radioactivity levels of sampling date were adjusted based on radioactive decay.

2.2.3.3   Estimation of Radiation Dose by Intra-body ^{137}Cs

Absorbed dose due to internal exposure inside the field mice that ingested ^{137}Cs was calculated using the Monte Carlo electron-photon transport code EGS5. The energy distribution and energy spectrum of β-particles and γ-photons of ^{137}Cs were written in the EGS5 user code based on ICRP Publication 107 [17]. The shape of the field mouse body was regarded as a cylinder. The size of the cylinder was decided based on the average of external measurements on five males and five females (a diameter of 20 mm and a height of 122 mm in male and a diameter of 17 mm and a height of 104 mm in female). Whole-body dose of the field mice were calculated from energy transfer to the cylinders assuming mice body. Energy transfer of β-particles or γ-photons to the tissues was calculated by simulating the reaction of ^{137}Cs randomly distributed in the cylinder corresponding to the whole body, from which β-particles or γ-photons were emitted in random directions.

EGS 5 assumes a body-equivalent substance, which composed of the same atoms and having the same density as the body tissue. EGS5 simulates the interaction with the body-equivalent substance and the β-particles or γ-photons at the atomic level. For all simulated interactions, EGS 5 calculated the energy transferred to the body-equivalent substances. EGS 5 calculates the energy (J) transferred to substances from the β-particles and γ-photons per unit volume by summing energy transferred to the body-equivalent substance within the unit volume (mm³). We calculated the energy transferred to the body-equivalent substance within the unit mass (J/kg) by dividing the energy transferred to the body-equivalent substance within the unit volume (J/mm³) by the density (kg/mm³) of the body-equivalent substance. We used the energy absorbed per unit mass (J/kg) as radiation dose (Gy) in the EGS5-simulation.

2.2.4   Total Dose Rate

Total dose rate was represented as the sum of the average of external dose rate and the average of internal dose rate in each year.

2.3   Results

2.3.1   External Dose Rate

The temporal change of the ambient dose equivalent rate of gamma-radiation at ground level of the equivalent trapping site during the study period from 2012 to 2016 is shown in Fig. 2.2. The highest average ambient equivalent dose rate, 16.7 ± 4.4 μSv/h, was observed in 2012. Then, the ambient dose equivalent rate gradually decreased over the study period.

Figure 2.3 shows the temporal change of external dose rate from 2012 to 2016. The highest average external dose rate, 0.335 ± 0.076 mGy/day was observed in 2012. The average of the external dose rate decreased gradually.

2.3.2   Internal Dose Rate

Table 2.1 displays the number of the field mice used to measure ^{137}Cs concentrations in this study. In total, 91 mice (male, 56; female, 35) were used. Cs-137 concentration in the field mice during the study period from 2012 to 2016 is shown in Fig. 2.4.

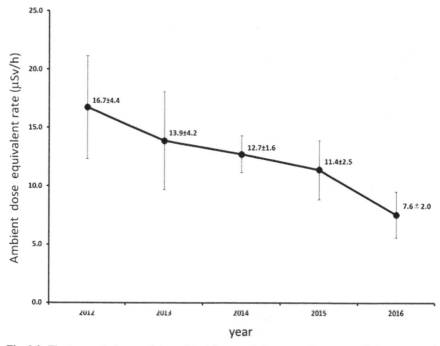

Fig. 2.2 The temporal change of the ambient dose equivalent rate of gamma-radiation at ground level of the trapping site

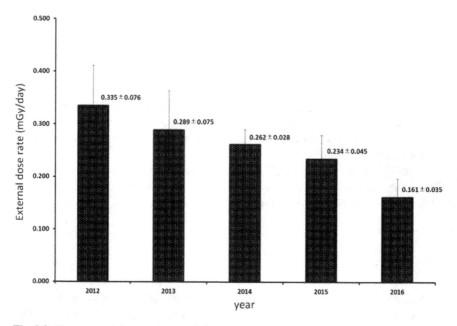

Fig. 2.3 The temporal change of external dose rate based on the ambient dose equivalent rate of gamma-radiation at ground level of the trapping site

Table 2.1 The number of the large Japanese field mouse (*Apodemus speciosus*) used to measure ^{137}Cs concentrations in the present study

Trapping time		Number		
Year	Month	Male	Female	Total in year
2012	Aug.–Nov.	15	12	27
2013	Jul.–Oct.	13	6	19
2014	Jun.–Aug.	10	6	16
2015	Aug.	8	3	11
2016	Aug.–Sep.	10	8	18
	Total	56	35	91

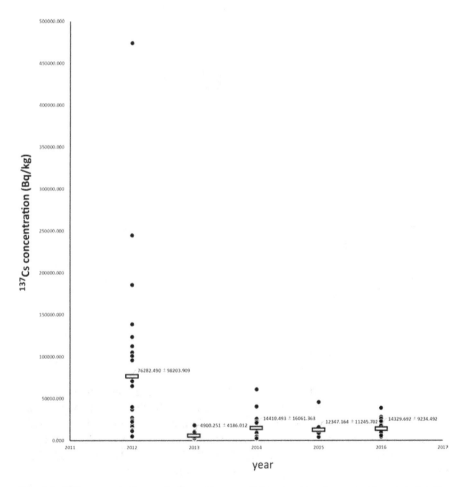

Fig. 2.4 ^{137}Cs concentration in the large Japanese field mice (*Apodemus speciosus*) during the study period from 2012 to 2016. ⬜: Average

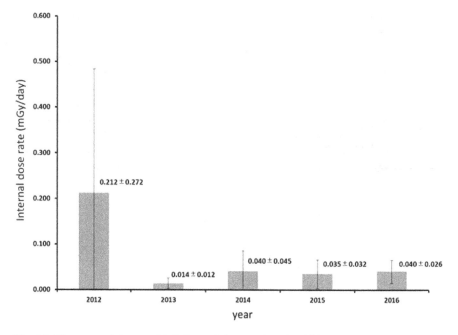

Fig. 2.5 The temporal change of daily internal dose rate based on ^{137}Cs concentrations in the large Japanese field mice body

The highest concentration, 474,076.1 Bq/kg, was observed in the female mice in 2012. Then, the concentration was indicated at less than 100,000 Bq/kg after 2013.

Figure 2.5 shows the temporal change of daily internal dose rate from 2012 to 2016. The average of the daily internal dose rate decreased sharply from 0.212 ± 0.272 to 0.014 ± 0.012 mGy/day between 2012 and 2013. Then, the value indicated approximately 0.040 mGy/day after 2014.

2.3.3   Total Dose Rate

The temporal change of the total dose rate of the trapping site during the study period from 2012 to 2016 is shown in Fig. 2.6. From 2012 to 2013, total dose rate decreased sharply from 0.547 to 0.303 mGy/day and then the value gradually decreased. The ratio of external dose rate to total dose rate was calculated to be 61.2%, 95.4%, 86.7%, 87.0% and 80.1% in 2012, 2013, 2014, 2015 and 2016, respectively (Fig. 2.7). Thus, main exposure was attributed to the external exposure during the study period.

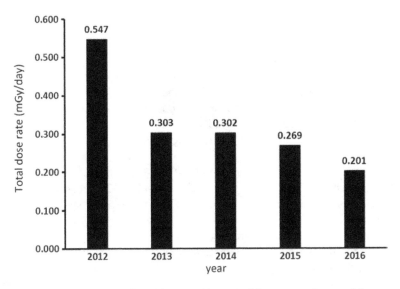

Fig. 2.6 The temporal change of total dose rate (the sum of the average of external dose rate and the average of internal dose rate)

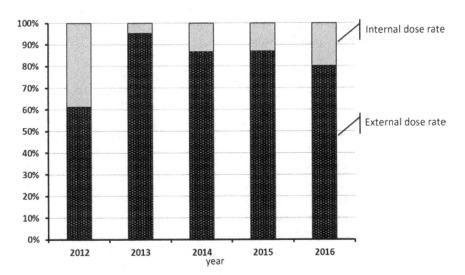

Fig. 2.7 The temporal change in the ratio of external dose rate and internal dose rate. The dose rate was mainly caused by the external dose rate during study period

2.4 Discussion

The purpose of this study was to determine whether the dose rate after 2012 caused reduced fertility in male and female field mice distributed in the DR zone. The external dose rate was calculated based on the ambient dose equivalent rate at ground level at the trapping sites. The internal dose rate was simulated using EGS5 based on ^{137}Cs concentrations in the captured field mice. Total dose rate (sum of external and internal dose rate) for the field mice within the DR zone was estimated to be 0.201–0.547 mGy/day.

According to the DCRLs determined by ICRP, exposure exceeding 1.0 mGy/day could reduce fertility in male and female field mice. Thus, it is concluded that the present dose rate of the field mice distributed in the trapping sites was less than the dose rate that reduces fertility. The result of the present study was supported by the report on male field mice reproductive researches conducted in the same trapping site [18]. No significant increase in the apoptotic cell frequencies nor the frequencies of morphologically abnormal sperm in the field mice captured in the DR zone (ambient dose equivalent rate ranging from 12.7 to 13.9 μSv/h) was observed as compared to the results of two control sites, Aomori Prefecture and Toyama Prefecture in 2013 and 2014. The resultant median values of the apoptotic cell frequency was 5.5–7.7 cells/mm^2 of cross section area in Fukushima, 8.0–9.2 in Aomori and 8.3–9.1 in Toyama. The results of the frequencies of morphologically abnormal sperm in 100 randomly selected spermatozoa showed that the abnormality was mainly observed on the midpiece of the sperm, with the median values of the frequencies being 12–18% in Fukushima, 13–19% in Aomori and 15–17% in Toyama. The statistical analysis of the frequency of apoptotic cells and that of morphologically abnormal sperm suggested no significant difference among the locations. It is reported that enhanced spermatogenesis was thought to have occurred in the field mice distributed in Tanashio (21 μGy/day), Murohara (304–365 μGy/day) and Akogi (407–447 μGy/day). The number of spermatogenic cells and proliferating cell nuclear antigen-positive cells per seminiferous tubule of the Akogi group was significantly higher than that in the Tanashio and Murohara groups. In addition, TUNEL-positive apoptotic cells tended to be detected at the lower level in the Murohara and Akogi groups compared to the Tanashio group where radiocontamination of the least among the three places [19].

The result of the present study suggested that the present dose rate of the field mice distributed in the trapping site does not affect the reproduction. However, it is difficult to generalize the result across the entire DR zone because levels of contamination by radioactive substances vary from location to location within the DR zone. The difference in levels of contamination may affect not only the external dose rate but also the internal dose rate through foraging activity. Therefore, further study, particularly on apoptotic cell frequencies and the frequencies of morphologically abnormal sperm, is required in heavily contaminated areas to compare the reported data. In addition, it is necessary to evaluate female reproductive organs,

Table 2.2 Potential biological effects by deferent level of radiation for Muridae species (ICRP 2018 [17])

Dose rate (mGy/day)	Effects
>1000	Mortality from haemopoietic syndrome in adults.
100–1000	Reduction in lifespan due to various causes.
10–100	Increased morbidity. Possible reduced lifespan. Reduced reproductive success.
1–10	Potential for reduced reproductive success due to reduced fertility in males and females.
0.1–1	Very low probability of effects. **Derived consideration reference level for Muridae.**
0.01–0.1	No observed effects.
<0.01	Natural background.

particularly the ovary, to gain an understanding of the actual situation of reproduction of the field mouse under the exposure. Furthermore, the effect of β-particles should be included to increase estimation accuracy of the external dose rate because the body of the field mouse is in contact with the contaminated ground in most of the time due to their activity pattern. It must be noticed that dose rate exceeding 0.1 mGy/day, which is mainly attributed to external dose rate is still prevalent in most of the DR zone. According to the DCRLs determined by ICRP, there is a very low probability of certain effects occurring under dose rate exceeding 0.1 mGy/day (Table 2.2) [11]. Thus, various indexes should be applied to evaluate the exposure effects on the species. In fact, chromosomal aberrations in the splenic lymphocytes of field mice captured near FNPP were reported [20]. Frequencies of translocations and dicentrics per cell (%) of the field mice distribution in "the heavily contaminated area (ambient dose equivalent rate = 80.0 μSv/h)" were significantly higher than that of the control area (ambient dose rate was 0.1 μSv/h), slightly contaminated area (ambient dose equivalent rate ranging from 0.3 to 0.4 μSv/h), and moderately contaminated area (ambient dose equivalent rate ranging from 7.5 to 30.7 μSv/h). Translocations and dicentrics were mainly observed in chromosome 2, potentially resulting from the presence of vulnerable sites to cellular stress in chromosome 2. These results, increased chromosomal aberrations, suggest that the present dose rate could affect the genome sequence of the species distributed in the DR zone. Investigation of the change of the mutation rate in germ cells is the first priority to evaluate the effect of the present dose rate on the field mouse population.

Acknowledgments We would like to thank Professor Hirayama for preparing the EGS5-code for estimating the internal exposure dose by ^{137}Cs. And thanks to Dr. Tsukasa Okano for the cooperation of the research.

References

1. Japan Meteorological Agency (2011) The 2011 Great East Japan earthquake. -Portal- http://www.jma.go.jp/jma/en/2011_Earthquake/2011_Earthquake.html. Accessed 28 Mar 2018
2. Nuclear and Industrial Safety Agency (2011) Discharge of radioactive materials to the environment (chapter VI). Report of Japanese government to the IAEA ministerial conference on nuclear safety
3. Hoshi H, Ogino M, Kawabe R, et al (2011) Computational analysis on accident progression of Fukushima Dai-ichi NPS. PSAM topical conference in Tokyo, Japan
4. Chino M, Nakayama H, Nagai H et al (2011) Preliminary estimation of release amounts of ^{131}I and ^{137}Cs accidentally discharged from the Fukushima Daiichi Nuclear Power Plant into the atmosphere. J Nucl Sci Technol 48:1129–1134
5. Terada H, Tatata G, Nagai H (2012) Atmospheric discharge and dispersion of radionuclides during the Fukushima Dai-ichi Nuclear Power Plant accident. Part II: verification of the source term and analysis of regional-scale atmospheric dispersion. J Environ Radioact 112:141–154
6. Sugimoto J (2014) Accident of Fukushima Daiichi nuclear power plant: sequences, fission products released, lessons learned. In: Takahashi S (ed) Radiation monitoring and dose estimation of the Fukushima nuclear accident. Springer, Tokyo
7. Imanaka T, Endo S, Sugai M et al (2012) Early radiation survey of Iitate village, which was heavily contaminated by the Fukushima Daiichi accident, conducted on 28 and 29 March 2011. Health Phys 102:680–686
8. National Center for Biotechnology Information (n.d.) PubChem compound database; CID=5486527. https://pubchem.ncbi.nlm.nih.gov/compound/5486527. Accessed 18 Mar 2018
9. Matsuda N, Mikami S, Shimoura S et al (2015) Depth profiles of radioactive cesium in soil using a scraper plate over a wide area surrounding the Fukushima Dai-ichi Nuclear Power Plant, Japan. J Environ Radioact 139:427–434
10. Takahashi J, Tamura K, Suda T et al (2015) Vertical distribution and temporal changes of 137Cs in soil profiles under various land uses after the Fukushima Dai-ichi Nuclear Power Plant accident. J Environ Radioact 139:351–361
11. ICRP (2008) Environmental protection: the concept and use of reference animals and plants, ICRP publication 108. Ann ICRP 38(4–6):1–242. (August–December 2008)
12. Garnier-Laplace J, Beaugelin-Seiller K, Hinton TG (2011) Fukushima wildlife dose reconstruction signals ecological consequences. Environ Sci Technol 45:5077–5078
13. Ohdachi SD, Ishibashi Y, Iwasa MA et al (2009) The wild mammals of Japan. Shoukadoh Book Sellers and the Mammal Society of Japan, Kyoto. 511 pp
14. Hirayam H, Namito Y, Bielajew AF, et al (2005) The EGS5 code system. KEK Report 2005-8, SLAC-R-730. High Energy Accelerator Research Organization (KEK), Stanford Linear Accelerator Center (SLAC). 2005, National Technical Information Service, U.S. Department of Commerce, Springfield, VA, USA
15. Committee of Reviewing Taxon Names and Specimen Collections, the Mammal Society of Japan (2009) Guideline for using wild mammals. Mamm Sci 49:303–319. (in Japanese)
16. Hikida T, Murakami O (1980) Age determination of the Japanese wood mouse, *Apodemus speciosus*. Nihon Seitai Gakkaishi 30:109–116. in Japanese
17. ICRP (2008) Nuclear decay data for dosimetric calculations. ICRP publication 107. Ann ICRP 38(3), Supplement p62
18. Okano T, Ishiniwa H, Onuma M et al (2016) Effects of environmental radiation on testes and spermatogenesis in wild large Japanese field mice (*Apodemus speciosus*) from Fukushima. Sci Rep. 23 6:23601

19. Takino S, Yamashiro H, Sugano Y et al (2017) Analysis of the effect of chronic and low-dose radiation exposure on spermatogenic cells of male Large Japanese Field Mice (*Apodemus speciosus*) after the Fukushima Daiichi Nuclear Power Plant Accident. Radiat Res 187:161–168
20. Kawagoshi T, Shiomi N, Takahashi H et al (2017) Chromosomal aberrations in Large Japanese Field Mice (*Apodemus speciosus*) captured near Fukushima Dai-ichi Nuclear Power Plant. Environ Sci Technol 51:4632–4641

3

FNPP Accident: Effect on Fishing Industry

Takami Morita, Daisuke Ambe, Shizuho Miki, Hideki Kaeriyama, and Yuya Shigenobu

Abstract On March 11, 2011, a massive earthquake and the resultant gigantic tsunami struck the Tohoku area of Japan (the Great East Japan Earthquake) and damaged many fishing boats and fishing ports. The earthquake and the subsequent tsunami also seriously caused the damage to Fukushima Daiichi Nuclear Power Plant (FNPP). Consequently, large amounts of radioactive cesium (Cs) were released into the atmosphere and ocean and subsequently polluted fishery products of Fukushima and adjacent prefectures. The Fukushima Prefectural Federation of Fisheries Cooperative Association (Fukushima FCA) has voluntarily stopped the fishing operations off Fukushima Prefecture since March 2011 due to the influence by the FNPP accident. The concentration of radioactive Cs in seawater rapidly decreased by ocean processes, and accordingly the concentration in fishery products did reduce. From April to June 2011, the proportion of inspected specimens off Fukushima exceeding the Japanese regulatory limit was 57.1%, but it decreased and continued to be 0% after April 2015. In addition, most of fishing industries in Fukushima Prefecture have been already restored from the damage by the earthquake and its aftermath. The Fukushima FCA started the trial fishing operation from June 2012.

Keywords Monitoring research · Fishery products · Radioactive cesium · Fukushima prefecture · Trial fishing operation

T. Morita (✉) · S. Miki · H. Kaeriyama · Y. Shigenobu
Radioecology Group, Research Center for Fisheries Oceanography and Marine Ecosystem, National Research Institute of Fisheries Science, Fisheries Research Agency, Yokohama, Japan
e-mail: takam@affrc.go.jp

D. Ambe
Fisheries Environment Group, Research Center for Fisheries Oceanography and Marine Ecosystem, National Research Institute of Fisheries Science, Fisheries Research Agency, Yokohama, Japan

3.1 Introduction

On March 11, 2011, a massive earthquake (moment magnitude 9.0) and a resultant gigantic tsunami struck the Tohoku area of Japan (the Great East Japan Earthquake). According to the report by the Fisheries Agency of Japan (FAJ), the aftermath of the earthquake damaged around 29,000 fishing boats and 319 fishing ports, which were approximately 10% of each total number in Japan [1]. Until January 2018, 18,614 fishing boats and all fishing ports have regained the function. The fishing industries in the tsunami-damaged areas excluding Fukushima Prefecture have been steadily recovering from the disaster. On the other side, the fishing industry in Fukushima has another unavoidable problem.

The earthquake and tsunami caused serious damage to FNPP. Consequentially, FNPP released a significant quantity of radionuclides into the atmosphere, and the fall out peaked around March 15, 2011 [2–4]. Although various radionuclides were released [5], the major radionuclides were radioactive iodine (I), ^{131}I (physical half-life; 8.02 days), and two kinds of radioactive cesium (Cs), ^{134}Cs (2.06 years) and ^{137}Cs (30.1 years). Radioactivity of ^{134}Cs and ^{137}Cs released was approximately equal [2]. The total quantity of ^{131}I and ^{137}Cs into the atmosphere between March 12, 2011, and May 1, 2011, was estimated to be approximately 200 PBq and 13 PBq, respectively [6]. Furthermore, the amount of ^{137}Cs deposited on the ocean surface from the atmosphere was estimated as 7.6 PBq [6] and 12–15 PBq [7], and meaning that most of ^{137}Cs released to the atmosphere was introduced into the ocean. In addition, extremely contaminated cooling water, which interacted with the ruptured nuclear fuel rods, was leaked from a cracked sidewall near the intake channel of Unit 2 reactor during April 1–6 in 2011. Nuclear Emergency Response Headquarters of Japan (NERH) estimated that the contaminated water contained 4.7 PBq of radionuclides including ^{131}I, ^{134}Cs, and ^{137}Cs [8]. Another report indicated that the direct release to the ocean had already been going on March 26, 2011, and estimated that the total amount of ^{137}Cs directly released was 3.5 ± 0.7 PBq from March 26, 2011 to the end of May 2011 [9].

The Fukushima FCA did not fully grasp the radioactive pollution of fishery products off Fukushima by the large amounts of radionuclides from FNPP in March 2011. Additionally, not all fishing boats and fishing ports in Fukushima Prefecture were damaged by the aftermath of the earthquake. However, the Fukushima FCA decided to voluntarily stop the fishing operations off Fukushima for the food safety on March 15, 2011 [10]. Consequently, this voluntary stop of the fishing operations has been continuing from March 2011 to the present (August, 2019), by the monthly update. In such a situation, the Fukushima FCA started the trial fishing operation from June 2012 off Fukushima [11].

In this chapter, we introduce the radioactive pollution in fishery products and the state of the fisheries industry in Fukushima Prefecture.

3.2   Radioactive Pollution of Fishery Products

Marine organisms generally incorporate [134]Cs and [137]Cs (radioactive Cs) both by uptake from seawater or by food ingestion. Therefore, radioactive Cs concentration in marine organisms depends strongly on that in surrounding seawater, which effects the concentration in the food organisms. The extremely highly contaminated water leaked directly into oceans around March to April in 2011 [8, 9]. However, as radioactive Cs concentration in seawater rapidly decreased by ocean processes [12], that in the pelagic fishes, invertebrates and seaweeds accordingly decreased [13–16]. Although it is pointed out that the decrease rate of radioactive Cs concentration in demersal fishes is slow [12], that in demersal fishes has been steadily decreasing over time in the monitoring research [17]. Figure 3.1 shows radioactive Cs concentration in rockfish (*Sebastes cheni*), one of demersal fishes, in both north and south of FNPP. The maximum concentration, 3200 Bq/kg-wet, was detected in the south area on July 6, 2011 [17]. Radioactive Cs concentration in the south area was higher than that in the north area, because extremely contaminated water mainly flowed to the south area [12, 18]. The traces of flow were found on the distribution map of radioactive Cs concentration in the sediment off Fukushima [19]. A previous research described that radionuclide bioavailability from contaminated sediment is typically low with the transfer factor [20] because of strong interaction of Cs mineral with clay minerals [21]. But, the slow decrease of radioactive Cs in demersal fishes was presumed as the result from continuing contamination of their food

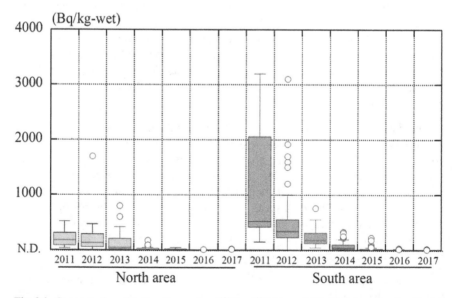

Fig. 3.1   Concentration of radioactive cesium ([134]Cs + [137]Cs) in rockfish (*Sebastes cheni*) collected in the north area (blue) and the south area (orange) from FNPP. N.D. shows not detected (detection limit < about 5.0 Bq/kg-wet for each radioactive Cs). (Data were cited from FAJ [17])

source (benthic infauna) from sediment [13, 15, 16, 22–25]. In fact, some studies after the FNPP accident indicated the presence of organically bound radioactive Cs, which can be bioavailable [26, 27].

The change of main year-class composing a population was a cause for decrease of radioactive Cs concentrations in fishery products including demersal fishes. A year-class-related differences of radioactive Cs concentration in some fish species, Pacific cod (*Gadus microcephalus*) and Japanese flounder (*Paralichthys olivaceus*), have been reported [28, 29]. The 2010 year-class and older classes had relatively higher concentrations of radioactive Cs than 2011 year-class and younger classes. The finding that year-classes born after the FNPP accident had lower concentration indicates that they incorporated only a small amount of radioactive Cs from the benthic food web.

High radioactive Cs concentration, 25,800 Bq/kg-wet, was detected in the specimen prepared from two greenlings (*Hexagrammos otakii*). The fish samples were caught close to Ota River approximately 20 km from FNPP in August 2012. Although these two outlier samples were thought to have migrated from the FNNP port, the probability of such outlier samples being found off Fukushima was exceedingly low [30]. Indeed, some fishes caught within the FNPP port remained highly contaminated, and the maximum concentration, 720,000 Bq/kg-wet, was detected in greenling collected on February 21, 2013. Therefore, TEPCO has set nets and prevented fishes from leaving the port. They also caught fishes in the FNPP port using fishing nets, and the total number of fishes was about 5000 samples from 2012 to 2018 [16, 30–32].

3.3   Monitoring Research in Fishery Products

The Japanese government has conducted the monitoring research of radionuclides (^{131}I, ^{134}Cs and ^{137}Cs) in fishery products for food safety since March 2011 (Fig. 3.2) [17]. TEPCO also carried out the monitoring research of fishery products collected in the 20-km zone from FNPP and in the FNPP port [33]. By the end of January 2018, over 50,000 fishery products off Fukushima have been inspected. Many independent research articles about fishery products have been published using these monitoring data and showed that radioactive Cs concentration in fishery products decreased [13–16, 22, 23, 32, 34]. The monitoring research by the Japanese government showed 57.1% of inspected fishery products off Fukushima were over the Japanese regulatory limit (100 Bq/kg-wet for radioactive Cs) in the period immediately following the accident (April–June 2011), but the ratio gradually reduced and continued to be 0% since April 2015 (Fig. 3.2). The statistical methods also demonstrated that the probability of occurrence of fishery products exceeding the Japanese regulatory limit was already extremely low in 2015 [35].

The Ministry of Health, Labour and Welfare (MHLW) of Japan also conducted the inspection of radioactive Cs in the food distributed on the market (Table 3.1) [36]. There were just two specimens over the Japanese regulatory limit (provisional

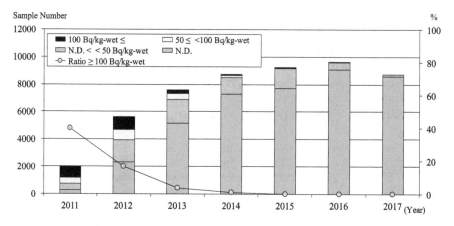

Fig. 3.2 Result of the monitoring research for radioactive cesium (^{134}Cs + ^{137}Cs) in fishery products collected off Fukushima. N.D.: not detected (detection limit < about 5.0 Bq/kg-wet for each radioactive Cs). (Data were cited from FAJ [17])

Table 3.1 Number of exceeding the Japanese regulatory limit in the inspection for distribution foods

Year[a]	Fishery products	Agricultural products	Meats (including wild animals)	Egg/dairy products	Others
2011	2 (0)[b]	57 (13)[b]	561 (91)[b]	0 (0)[b]	91 (18)[b]
2012	2	6	1	0	11
2013	0	8	0	0	2
2014	0	8	0	0	2
2015	0	8	1	0	3
2016	0	9	0	0	2
2017[c]	0	9	0	0	4

Data were cited from MHLW [36]
[a]Fiscal year from April to March in the next year
[b]Numbers in parentheses are numbers exceeding provisional regulation limit (500 Bq/kg-wet)
[c]April in 2017 to January in 2018

regulation limit in 2011) in fishery products since 2011. The FAJ succeeded in preventing the contaminated fishery products from being distributed. They also properly stopped the distribution of contaminated feed for aquaculture. Therefore, none of the aquaculture fishery products exceeded the Japanese regulatory limit (provisional regulation limit in 2011) not only in distribution food inspection but also in the monitoring research.

Radioactive strontium (^{89}Sr and ^{90}Sr) is one of the nuclear fission products as well as ^{137}Cs. Therefore, many people have been concerned about pollution of fishery products by radioactive Sr since the FNPP accident happened. Strontium-90 had been detected in fishery products before the FNPP accident [37]. The main source of ^{90}Sr in the North Pacific Ocean off Japan was the global and close-in radioactive fallout after the atmospheric nuclear bomb tests and the Chernobyl NPP accident [38, 39]. The average concentration of ^{90}Sr in marine fishes of the North Pacific Ocean off Japan was 0.025 ± 0.021 Bq/kg-wet in the past two decades before the FNPP accident [40].

Owing to the lower volatility than Cs, the amount of ^{90}Sr released into the atmosphere from the FNPP accident is estimated to be about 0.14 PBq [41], which is two orders of magnitude smaller than that of radioactive ^{137}Cs, 20 PBq [6, 7]. An extremely contaminated cooling water was leaked from a cracked sidewall near the intake channel of Unit 2 from late March to early April 2011 [8, 9]. The contaminated water including high concentration of radioactive Sr and Cs directly flowed into the ocean. Based on the initial ^{137}Cs/^{90}Sr activity ratio released from the FNPP accident and 3.5 PBq of ^{137}Cs directly was released into the ocean, the amount of ^{90}Sr in the contaminated water was estimated to be approximately 0.04 PBq [9, 41]. The International Atomic Energy Agency (IAEA) summarized the concentration factor of many elements for various aquatic organisms [42]. The value of concentration factor for Sr in marine fishes was 3 and lower than that for Cs, 100. In addition, ^{90}Sr in seawater was immediately diluted to the background level [41]. Therefore, ^{90}Sr concentration in fishery products was notably lower than that of ^{137}Cs even off Fukushima, and ^{90}Sr derived from the FNPP accident would not be detected in fishery products caught outside off Fukushima [40]. In conclusion, the influence of the FNPP accident by ^{90}Sr pollution on fishery products has been limited to the area off Fukushima though negligible [40, 43].

3.4 Trial Fishing Operations in Fukushima

The fishery products off Fukushima were polluted by large amounts of radioactive Cs released from FNPP. However, the Japanese and the Fukushima prefectural governments did not revoke the fishing licenses in Fukushima Prefecture, because the Fukushima FCA voluntarily continued to stop the fishing operations off Fukushima for the food safety from March 15, 2011 by the monthly update. In February 2012, the Fukushima FCA established the Fukushima Prefectural Fisheries Reconstruction Committee (Fukushima FRC) with external experts in order to reconstruct the fishery industry and restart the fishing operation off Fukushima [10].

In June 2012, based on the advice of the Fukushima FRC, the Fukushima FCA decided to carry out the trial fishing operation off Fukushima. The trial fishing operation has several limitations: the target species, the fishing methods, the number of days a week for operation, landing ports, the amount of landed fishes, the number of vessels involved in operations and the areas for operations [10]. The areas for the trial fishing operations are shown in Fig. 3.3. At first, the trial fishing operation for bottom trawling was performed only in area 1. The operation area has been expanded to include area 2 in October 2012, area 3 in February 2013, area 4 in May 2013, area 5 in August 2013, area 6 in December 2013, area 7 in October

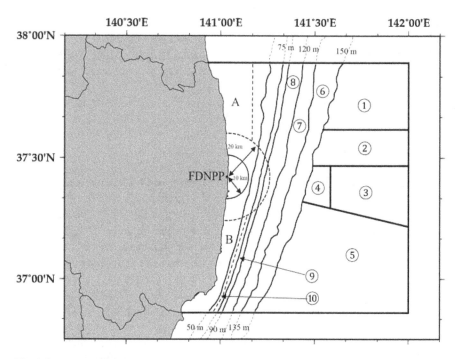

Fig. 3.3 Areas for the trial fishing operation. (Data were cited from Fukushima FCA [11])

2014, area 8 excluding the 20-km zone in October 2015, area 9 in February 2017 and area 10 in October 2017. The trial fishing operations of shipping net for young sand lance (*Ammodytes personatus*) and piercing net for icefish (*Salangichthys. ishikawae*) were conducted in areas A and B surrounded by the 20-km zone line (dotted lines) and land, respectively (Fig. 3.3). The trial fishing operation of shipping net in areas A and B was allowed in March 2013 and February 2014, respectively. The trial fishing operation of piercing net in areas A and B was done in February 2014. The range of the trial fishing operation did not include the 20-km zone until the end of February 2017. From March 2017, the range of the trial fishing operation covered all the areas excluding the 10-km zone off Fukushima. The reason for excluding the10-km zone was the presence of the fishes having high concentration of radioactive Cs in the FNPP port. As described above, TEPCO has set the net preventing fishes from coming in and going out the port and also exterminated fishes in the port by fishing nets [33].

The total number of target species for the trial fishing operation gradually increased (Fig. 3.4). First target species were three invertebrate species, two kinds of octopuses (*Enteroctopus dofleini* and *Octopus conispadiceus*), and a whelk (*Buccinum isaotakii*). These species were selected based on the monitoring research result, which showed no detection of radioactive Cs in these species (detection limit is under about 5.0 Bq/kg-wet for each radioactive Cs). The monitoring result was consistent with the previous report that concentration factor of

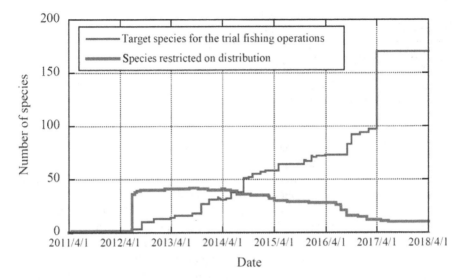

Fig. 3.4 Temporal change of the number of target species for the trial fishing operation (blue line) and the species restricted on the distribution (red line). (Data were cited from FAJ [17] and Fukushima FCA [11])

radioactive Cs in marine mollusc species is lower, up to 9 for cephalopods and 60 for others as compared with marine fish species, up to 100 [42]. In April 2018, the Fukushima FCA recognized all fish species for fishery industry off Fukushima as the target species except for 10 species restricted on the distribution (about 170 species). The ten species were surfperch (*Ditrema temminckii),* fox jacopever (*Sebastes vulpes*), Japanese black seabream (*Acanthopagrus schlegelii*), masu salmon (*Oncorhynchus masou*), rockfish (*Sebastes cheni*), Japanese sea bass (*Lateolabrax japonicus*), starry flounder (*Platichthys stellatus*), Spotbelly rockfish (*Sebastes pachycephalus*), marbled rockfish (*Sebastiscus marmoratus*) and Stimpson's hard clam (*Mercenaria stimpsoni*). However, radioactive Cs concentration in these species was not necessarily high even in March 2018. The respective concentration for ten species has been below 25 Bq/kg-wet from October 2017 to March 2018 [17]. The Japanese government had not restricted the fishery products except for young sand lance off Fukushima until launching the trial fishing operation, because the fishing operation off Fukushima was voluntarily stopped by the Fukushima FCA. When the trial fishing operation off Fukushima started in June 2012, 36 species were restricted on the distribution by the government. The numbers of species restricted on the distribution reached the maximum of 42 species in August 2013 and gradually decreased to 10 species in June 2017 (Fig. 3.4). The restriction of the distribution is lifted by the Japanese Government, when the result of monitoring research is stably and sufficiently below the Japanese regulatory limit. However, the criteria for lifting the distribution restriction has not yet been announced. The development of scientific/statistical framework for lifting the distribution restriction will be required in order to promptly restore the livelihoods of local farmers, fishermen, and other food producers [44].

Table 3.2 Amount of landing in the trial fishing operations in Fukushima Prefecture and comparison with that before the FNPP accident

Year	2010	2012	2013	2014	2015	2016	2017
Amount of landing (t)	25,914	122	406	742	1512	2100	3286
Comparison with 2010 (%)	–	0.471	1.57	2.86	5.83	8.10	12.7

Data were provided from the Fukushima FCA

The catch amount of the trial fishing operation has been gradually increasing year by year, but the amount in 2017 was 12.7% of that before the FNPP accident (Table 3.2). Although most of the fishing boats and the ports in Fukushima Prefecture have already been recovered, the increase of catch amount has been slow. The reason is that the FNPP accident has lost the confidence of consumer and distributor for the safety of fishery products off Fukushima, and additionally most of them could not purchase the Fukushima fishery products due to the reduced distribution amount, so, they switched to other production areas for purchasing fishery products. In order to restore their lost confidence, it is important to continue careful monitoring research and provide scientific information supporting the monitoring research results. We hope the distribution amount off Fukushima will increase in the future.

Acknowledgments We appreciate the staff members of the Fukushima FCA for providing the amount of landing data in the trial fishing operation. We thank all fishery workers in Fukushima Prefecture for cooperating with the monitoring research. We also thank the staff members of Radioecology Group for great help.

References

1. FAJ (2018a) The great East Japan earthquake's impact on fisheries and future measures. http://www.jfa.maff.go.jp/attach/pdf/index-47.pdf. Accessed 24 Mar 2018. (in Japanese)
2. Chino M, Nakayama H, Nagai H et al (2011) Preliminary estimation of release amounts of ^{131}I and ^{137}Cs accidentally discharged from the Fukushima Daiichi Nuclear Power Plant into the atmosphere. J Nucl Sci Technol 48:1129–1134
3. Morino Y, Ohara T, Nishizawa M (2011) Atmospheric behavior, deposition, and budget of radioactive materials from the Fukushima Daiichi Nuclear Power Plant in March 2011. Geophys Res Lett 38:L00G11
4. Masumoto Y, Miyazawa Y, Tsumune D et al (2012) Oceanic dispersion simulations of ^{137}Cs released from the Fukushima Daiichi Nuclear Power Plant. Elements 8:207–212
5. Steinhauser G (2014) Fukushima's forgotten radionuclides: a review of the under- studied radioactive emissions. Environ Sci Technol 48:4649–4663
6. Kobayashi T, Nagai H, Chino M et al (2013) Source term estimation of atmospheric release due to the Fukushima Dai-ichi Nuclear Power Plant accident by atmospheric and oceanic dispersion simulations. J Nucl Sci Technol 50:255–264
7. Aoyama M, Kajino M, Tanaka TY et al (2016) ^{134}Cs and ^{137}Cs in the North Pacific Ocean derived from the TEPCO Fukushima Dai-ichi Nuclear Power Plant accident, Japan in March 2011: part two – estimation of ^{134}Cs and ^{137}Cs inventories in the North Pacific Ocean. J Oceanogr 72:67–76

8. NERH (2011) Report of the Japanese government to the IAEA ministerial conference on nuclear safety. http://www.kantei.go.jp/foreign/kan/topics/201106/iaea_houkokusho_e.html. Accessed 27 Jan 2014
9. Tsumune D, Tsubono T, Aoyama M et al (2012) Distribution of oceanic ^{137}Cs from the Fukushima Daiichi nuclear power plant simulated numerically by a regional ocean model. J Environ Radioact 111:100–108
10. Yagi N (2016) Impact of the nuclear power plant accident and the start of trial operations in Fukushima fisheries. In: Nakanishi TM, Tanoi T (eds) Agricultural implications of the Fukushima nuclear accident. Springer, Tokyo, pp 217–228
11. Fukushima FCA (2018) The trial fishing operations off Fukushima. http://www.fsgyoren.jf-net.ne.jp/siso/sisotop.html. Accessed Feb 2014. (in Japanese)
12. Aoyama M, Tsumune D, Uematsu M et al (2012) Temporal variation of [134]Cs and ^{137}Cs activities in surface water at stations along the coastline near the Fukushima Dai-ichi Nuclear Power Plant accident site, Japan. Geochem J 46:321–325
13. Buesseler KO (2012) Fishing for answers off Fukushima. Science 338:480–482
14. Iwata K, Tagami K, Uchida S (2013) Ecological half-lives of radiocesium in 16 species in marine biota after the TEPCO's Fukushima Daiichi nuclear power plant accident. Environ Sci Technol 47:7696–7703
15. Wada T, Nemoto Y, Shimamura S et al (2013) Effects of the nuclear disaster on marine products in Fukushima. J Environ Radioact 124:246–254
16. Wada T, Fujita T, Nemoto Y et al (2016) Effects of the nuclear disaster on marine products in Fukushima: an update after five years. J Environ Radioact 164:312–324
17. FAJ (2018b) Results of the monitoring on radioactivity level in fisheries products. http://www.jfa.maff.go.jp/e/inspection/index.html. Accessed 24 Mar 2018
18. Tsumune D, Tsubono T, Aoyama M et al (2013) One-year, regional-scale simulation of ^{137}Cs radioactivity in the ocean following the Fukushima Dai-ichi Nuclear Power Plant accident. Biogeosciences 10:5601–5617
19. Ambe D, Kaeriyama H, Shigenobu Y et al (2014) Five-minute resolved spatial distribution of radiocesium in sea sediment derived from the Fukushima Dai-Ichi nuclear power plant. J Environ Radioact 138:264–275
20. Fowler SW, Fisher NS (2004) Radionuclides in the biosphere. In: Livingston HD (ed) Marine radioactivity. Elsevier, Amsterdam, pp 167–203
21. Comans RNJ, Hockley D (1992) Kinetics of cesium sorption on illite. Geochim Cosmochim Acta 56:1157–1164
22. Tateda Y, Tsumune D, Tsubono T (2013) Simulation of radioactive cesium transfer in the southern Fukushima coastal biota using a dynamic food chain transfer model. J Environ Radioact 124:1–12
23. Tateda Y, Tsumune D, Tsubono T et al (2015) Radiocesium biokinetics in olive flounder inhabiting the Fukushima accident-affected Pacific coastal waters of eastern Japan. J Environ Radioact 147:130–141
24. Sohtome T, Wada T, Mizuno T et al (2014) Radiological impact of TEPCO's Fukushima Dai-Ichi nuclear power plant accident on invertebrates in the coastal benthic food web. J Environ Radioact 138:106–115
25. Wang C, Baumann Z, Daniel J et al (2016) Contaminated marine sediments as a source of cesium radioisotopes for benthic fauna near Fukushima. Environ Sci Technol 50:10448–10455
26. Otosaka S, Kobayashi T (2013) Sedimentation and remobilization of radiocesium in the coastal area of Ibaraki, 70 km south of the Fukushima Dai-Ichi nuclear power plant. Environ Monit Assess 185:5419–5433
27. Ono T, Ambe D, Kaeriyama H et al (2015) Concentration of [134]Cs + [137]Cs bonded to the organic fraction of sediments offshore Fukushima, Japan. Geochem J 49:219–227
28. Kurita Y, Shigenobu Y, Sakuma T et al (2015) Radiocesium contamination histories of Japanese flounder (*Paralichthys olivaceus*) after the 2011 Fukushima Nuclear Power Plant accident. In: Nakata K, Sugisaki H (eds) Impact of the Fukushima nuclear accident on fish and fishing grounds. Springer, Tokyo, pp 139–154

29. Narimatsu Y, Sohtome T, Yamada M et al (2015) Why do the radionuclide concentrations of Pacific cod depend on the body size? In: Nakata K, Sugisaki H (eds) Impact of the Fukushima nuclear accident on fish and fishing grounds. Springer, Tokyo, pp 123–138

30. Shigenobu Y, Fujimoto K, Ambe D et al (2014) Radiocesium contamination of greenlings (*Hexagrammos Otakii*) off the coast of Fukushima. Sci Rep 4:6851

31. Fujimoto K, Miki S, Kaeriyama H et al (2015) Use of otolith for detecting strontium- 90 in fish from the harbor of Fukushima Dai-Ichi nuclear power plant. Environ Sci Technol 49:7294–7301

32. Tateda Y, Tsumune D, Misumi K et al (2017) Biokinetics of radiocesium depuration in marine fish inhabiting the vicinity of the Fukushima Dai-ichi Nuclear Power Plant. J Environ Radioact 166:67–73

33. TEPCO (2017) Analysis results of fish and shellfish (the ocean area within 20km radius of Fukushima Daiichi NPS). http://www.tepco.co.jp/en/nu/fukushima-np/f1/smp/index-e.html. Accessed 26 Mar 2018

34. Buesseler K, Dai M, Aoyama M et al (2017) Fukushima Daiichi-derived radionuclides in the ocean: transport, fate, and impacts. Annu Rev Mar Sci 9:173–203

35. Okamura H, Ikeda S, Morita T et al (2016) Risk assessment of radioisotope contamination for aquatic living resources in and around Japan. Proc Natl Acad Sci U S A 113:3838–3843

36. MHLW (2018) Levels of radioactive contaminants in foods tested in respective prefectures

37. Morita T, Fujimoto K, Kasai H et al (2010) Temporal variations of ^{90}Sr and ^{137}Cs concentrations and the ^{137}Cs/^{90}Sr activity ratio in marine brown algae, *Undaria pinnatifida* and *Laminaria longissima*, collected in coastal areas of Japan. J Environ Monit 12:1179–1186

38. Bowen VT, Noshkin VE, Livingston HD et al (1980) Fallout radionuclides in the Pacific Ocean: vertical and horizontal distributions, largely from GEOSECS stations. Earth Planet Sci Lett 49:411–434

39. UNSCEAR (United Nations Scientific Committee on the Effects of Atomic Radiation) (2000) Exposure and effects of the Chernobyl accident (annex J). United Nations, New York

40. Miki S, Fujimoto K, Shigenobu Y et al (2017) Concentrations of ^{90}Sr and ^{137}Cs/^{90}Sr activity ratios in marine fishes after the Fukushima Dai-ichi Nuclear Power Plant accident. Fish Oceanogr 26:221–233

41. Povinec PP, Hirose K, Aoyama M (2012) Radiostrontium in the western North Pacific: characteristics, behavior, and the Fukushima impact. Environ Sci Technol 46:10356–10363

42. IAEA (2004) Sediment distribution coefficients and concentration factors for biota in the marine environment, Technical reports series, vol 422. International Atomic Energy Agency, Vienna

43. Karube Z, Inuzuka Y, Tanaka A et al (2016) Radiostrontium monitoring of bivalves from the Pacific coast of eastern Japan. Environ Sci Pollut Res Int 23:17095–17104

44. Matsuzaki SS, Kumagai NH, Hayashi T (2016) Need for systematic statistical tools for decision-making in radioactively contaminated areas. Environ Sci Technol 50:1075–1076

4

Impact of Contamination due to Radioactive Cesium on Detritivores and Arthropods

Sota Tanaka, Tarô Adati, Tomoyuki Takahashi, and Sentaro Takahashi

Abstract To understand the behavior of radioactive cesium (^{134}Cs+^{137}Cs) in terrestrial invertebrates, chronological changes in the concentration of radioactive Cs in arthropods from different trophic levels were investigated after the Fukushima Daiichi Nuclear Power Plant accident. In addition, the level of radioactive Cs in earthworms, representing the detritivores, was also investigated. The median radioactive Cs concentration in the rice grasshopper (*Oxya yezoensis*) and the Emma field cricket (*Teleogryllus emma*) was 0.46 and 0.15 Bq/g fresh weight (fw) in 2012, respectively, which dropped continuously to 0.05 and 0.01 Bq/g fw in 2016. In contrast, no significant reduction in radioactive Cs concentration was observed in the Jorô spider (*Nephila clavata*) in which the concentration was 0.31 Bq/g fw in 2012 and remained at 0.14 Bq/g fw in 2016. The comparison of radioactive Cs concentration at each trophic level showed that the amount in detritivorous earthworms was 85 times higher than in herbivorous grasshoppers. This suggests that detritus food web could be a primary pathway for migration of radioactive Cs through food webs.

S. Tanaka (✉)
Kyoto University Graduate School of Agriculture, Kyoto, Japan

Research Group for Environmental Science, Nuclear Science and Engineering Center,
Japan Atomic Energy Agency, Ibaraki, Japan
e-mail: tanaka.sota.57s@st.kyoto-u.ac.jp; tanaka.sota@jaea.go.jp

T. Adati
Department of International Agricultural Development, Faculty of International Agriculture
and Food Studies, Tokyo University of Agriculture, Tokyo, Japan

T. Takahashi
Kyoto University Graduate School of Agriculture, Kyoto, Japan

Division of Radiation Control, Institute for Integrated Radiation and Nuclear Science,
Kyoto University, Kumatori, Sennan, Osaka, Japan

S. Takahashi
Division of Radiation Control, Institute for Integrated Radiation and Nuclear Science,
Kyoto University, Kumatori, Sennan, Osaka, Japan

Keywords Grasshopper · Cricket · Spider · Earthworm · Radioactive cesium

4.1 Introduction

The accident at TEPCO's Fukushima Daiichi Nuclear Power Plant (FNPP) caused serious radioactive contamination across a wide area of eastern Japan. There is concern about the effect on the environment of cesium-134 (^{134}Cs) and Cs-137 (^{137}Cs), with the relatively long half-life (^{134}Cs, 2.06 years; ^{137}Cs, 30.17 years). In Fukushima Prefecture, approximately 71% of the total land area is covered with forests [1]. Therefore, a large proportion of the radioactive Cs (^{134}Cs+^{137}Cs) released into the atmosphere was deposited onto the forested areas. Over time, the radioactive Cs accumulates in the soil surface layer and is maintained long-term in the soil organic layer. Thus, the organic layer of soil represents a major pool of radioactive Cs in forest ecosystems [2]. Radioactive Cs can then circulate throughout the forest ecosystem by biological processes [3] as is in the same elemental family as potassium, an essential element for all organisms.

Arthropods have a large biomass and are important food sources for other organisms such as birds, amphibians, reptiles and mammals. Terrestrial arthropods are also important seasonal diets for freshwater fishes such as trout [4, 5] and act as a trophic linkage across the forest–stream ecosystem [6]. This supports the hypothesis that terrestrial arthropods could be a carrier of radioactive Cs throughout the food web and a route of transfer or radioactive Cs between the forest and aquatic ecosystems.

Earthworms function as ecosystem engineers [7] to produce a homogenized organic soil layer. This bioturbation activity is an important factor in the long-term behavior of radioactive Cs in the soil [8]. Earthworms are also important food resources for various organisms. Therefore, determining radioactive Cs concentration in arthropods and earthworms is important for understanding the long-term behavior of radioactive Cs in the ecosystem and for radiation risk assessment for non-human species. However, the number of reports on radioactive contamination in arthropods and earthworms after the FNPP accident is limited (Table 4.1) [9–19].

We reported the chronological changes in radioactive Cs levels in common terrestrial arthropod species from different trophic levels after the FNPP accident. The report showed a continuous annual reduction of radioactive Cs in herbivorous grasshoppers and omnivorous crickets, in contrast to carnivorous spiders in which no significant reduction in radioactive Cs was observed from 2012 to 2014. These differences in radioactive Cs at each trophic level over time suggest that the level of contamination of the varied food resource pathways is different [15].

The present study shows the latest data on the chronological changes of radioactive Cs concentration in the arthropods over the 5-year period from 2012 to 2016. Furthermore, in order to investigate the detritus food web which has been suggested to act a major transfer pathway of radioactive Cs in the food web [10, 16], detritivorous earthworms were investigated in 2014, and radioactive Cs concentration among species with different feeding habits was compared.

Table 4.1 Reports on radionuclide contamination in arthropods and earthworms after the FNPP accident

Target	Radionuclide	Collection year	References
Spider	^{134}Cs,^{137}Cs	2012	Ayabe et al. [9]
Various organisms including arthropods	^{137}Cs	2012	Murakami et al. [10]
Aquatic insects in a stream	^{134}Cs,^{137}Cs	2012–2013	Yoshimura and Akama [11]
Various organisms including spider	^{134}Cs,^{137}Cs, ^{110m}Ag	2011–2014	Nakanishi et al. [12]
Spider	^{134}Cs,^{137}Cs	2012–2013	Ayabe et al. [13]
Various organisms including arthropods	^{137}Cs	2012–2013	Sakai et al. [14]
Spider, cricket, grasshopper	^{134}Cs,^{137}Cs	2012–2014	Tanaka et al. [15]
Forest insects	^{137}Cs	2012–2013	Ishii et al. [16]
Earthworm	^{134}Cs,^{137}Cs	2011	Hasegawa et al. [17]
Earthworm	^{134}Cs,^{137}Cs	2011–2013	Hasegawa et al. [18]
Earthworm	^{137}Cs	2014–2016	Tanaka et al. [19]

4.2 Materials and Methods

4.2.1 Sampling Site and Measurement of Ambient Dose Equivalent Rate

The sampling site is located 40.1 km northwest of FNPP (latitude, 37°41'35" N; longitude, 140°44'08" E; Fig. 4.1). This area is composed of a hilly and mountainous landscape with agricultural fields and residential areas surrounded by mountainous forests. Residents were still not permitted to live in this area during the study period from 2012 to 2016.

Ambient dose equivalent rate was measured at multiple points on the sampling site using a NaI scintillation survey meter (TCS171, Hitachi, Ltd., Japan) placed 1 m above the ground, and the distance of each measurement point was at least 20 m.

4.2.2 Sampling of Arthropods and Earthworms

Three arthropod species, the rice grasshopper (*Oxya yezoensis*), the Emma field cricket (*Teleogryllus emma*) and the Jorô spider (*Nephila clavata*) were collected by sweep-net sampling and hand collection from September to October of each year between 2012 and 2016. For each arthropod species, we collected 20–200 individuals each year. Epigeic earthworms were collected by hand collection in October

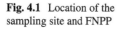

Fig. 4.1 Location of the sampling site and FNPP

2014. All of the sampled arthropods and earthworms experience a generational turn-over once a year.

4.2.3 Measurement of Radioactive Cs Levels in Samples

Gamma-ray spectrometry was conducted using a high-purity germanium detector (GC-2020, Canberra Industries, and GEM30-70, ORTEC, USA) with a multichannel analyzer (MCA, DAS1000, Canberra Industries, and Easy-MCA-8k, ORTEC, USA). The counting efficiency of the detector was determined by measuring a certified mixed radioactive standard gamma volume source (MX033SPLU8, Japan Radioisotope Association, and 24FY039, Japan Chemical Analysis Center). All samples were placed into 100 ml plastic containers (U-8), and the radioactivity of the samples was obtained as Bq/g fresh weight (Bq/g fw).

4.2.4 Statistical Analyses

For changes in ambient dose equivalent rate and radioactive Cs concentration in arthropods and earthworms, the lower (Q_1) and upper (Q_3) quartiles and the inter-quartile range (IQR = Q_3-Q_1) were calculated. Differences in the values among years were analyzed with a Kruskal–Wallis test using R version 2.15.3 [20].

4.3 Results

4.3.1 Ambient Dose Equivalent Rates at the Sampling Site

Ambient dose equivalent rate at the sampling site is shown in Fig. 4.2. The median ambient dose equivalent rate showed significant reductions from 3.74 to 1.29 μSv/h between 2012 and 2016 (Kruskal–Wallis test, $P < 0.05$). The decrease during the initial phase of the survey period from 2012 to 2013 was 29%, and the later phase from 2015 to 2016 was only 4%. The total reduction in the median ambient dose equivalent rate during the survey period from 2012 to 2016 was calculated to be 65%.

4.3.2 Chronological Changes in Radioactive Cs Levels in Arthropods

The change in radioactive Cs concentration in three arthropod species over the 5-year investigation is shown in Fig. 4.3. The median concentration of radioactive Cs in grasshoppers significantly decreased from 0.46 to 0.05 Bq/g fw between 2012 and 2016 (Kruskal–Wallis test, $P < 0.05$; Fig. 4.3A). Field crickets also showed a significant decrease in radioactive Cs concentration from 0.15 to 0.01 Bq/g fw ($P < 0.05$; Fig. 4.3B). In contrast, the median concentration of radioactive Cs in Jorô spiders was not significantly different during the survey period ($P = 0.14$; Fig. 4.3C); the median of radioactive Cs concentration from 2012 to 2016 were 0.31, 0.33, 0.20, 0.23, and 0.14 Bq/g fw, respectively. The decrease in median radioactive Cs

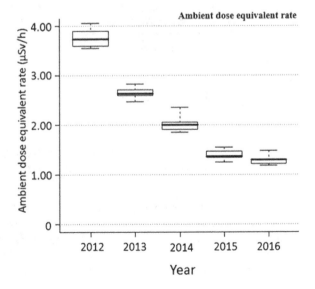

Fig. 4.2 The ambient dose equivalent rate (μSv/h at 1 m above ground) significantly differed over time after the FNPP accident (Kruskal-Wallis test, $P < 0.001$). Minimum and maximum values of dose rate are depicted by whiskers plots. The box signifies the upper and lower quartiles, and the median is represented by a horizontal line within the box for each year

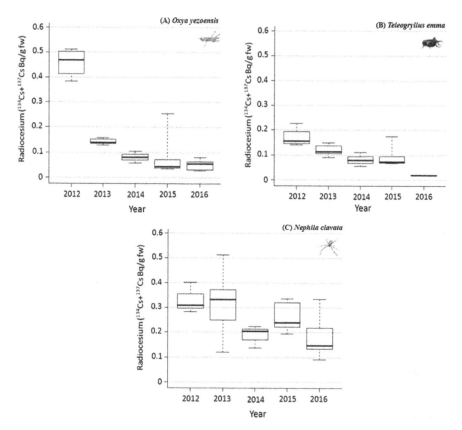

Fig. 4.3 Radioactive Cs concentration (Bq/g fw) in the rice grasshopper (*Oxya yezoensis*) (**A**), and the Emma field cricket (*Teleogryllus emma*) (**B**), significantly changed over time after the FNPP accident (Kruskal–Wallis test, (**A**) P = 0.002; (**B**) P = 0.003). However, no significant changes were observed in the Jorô spider (*Nephila clavata*) (**C**: P = 0.149). Minimum and maximum concentrations are depicted by whisker plots. The box signifies the upper and lower quartiles, and the median is represented by a horizontal line within the box for each year

concentration from 2012 to 2016 was calculated to be 88%, 87%, and 52% for grasshoppers, field crickets, and Jorô spiders, respectively.

4.3.3 Comparison Between Different Feeding Habits and ^{137}Cs Concentration

Radioactive Cs concentration in grasshoppers, field crickets, Jorô spiders and earthworms was compared in 2014. The median of ^{137}Cs concentration in earthworms was 4.87 Bq/g fw, which was about 85 times higher than that of grasshoppers, and over 30 times higher than that of Jorô spiders, which showed the highest level of radioactive Cs among the three arthropod species examined (Fig. 4.4).

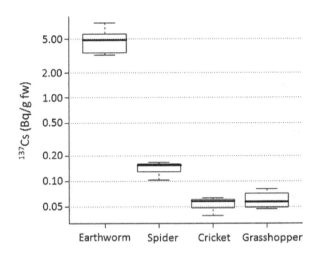

Fig. 4.4 Comparison of radioactive Cs concentration between different feeding habits in 2014

4.4 Discussion

The ambient dose equivalent rate at the sampling site significantly decreased from 2012 to 2016. The γ-ray contribution of ^{134}Cs to ambient dose equivalent rate is higher than ^{137}Cs [21], and the physical half-life of ^{134}Cs is shorter than that of ^{137}Cs. Therefore, the reduction in the ambient equivalent dose rate during the initial phase after the accident determined in the present study is mainly attributed to the physical half-life of ^{134}Cs, although the weathering effect may also contribute to the decrease. From 2015 to 2016, the decrease in ambient equivalent dose rate at the sampling site was only 4%, which was lower than the 29% decrease during the initial phase from 2012 to 2013. This suggests that the yearly rate of decrease in the ambient dose equivalent rate at the time of this study will remain for a long time even at the low levels.

The median radioactive Cs concentration in the grasshoppers and field cricket significantly decreased from 2012 to 2016. In contrast, the concentration in Jorô spiders did not differ significantly during the survey period. These trends are consistent with our previous report from 2012 to 2014 [15]. The radioactive Cs concentration in Jorô spiders showed relatively wide variation during the 5-year sampling period, and the median radioactive Cs concentration in Jorô spiders remained at 52% between 2012 and 2016. Different food resource pathways, such as grazing and detrital food web, likely caused this trend [22]. As herbivorous, grasshoppers rely on the grazing food web, and their radioactive Cs concentration showed a rapid decrease. These indicate that radioactive Cs transfer from the grazing food web is not a dominant long-term contributor in grasshoppers. In contrast, carnivorous web spiders rely on both the grazing and detritus food webs due to the variety of prey items captured by their orb web. The variation in radioactive Cs in Jorô spiders during the 5-year survey period indicated that they use both the grazing and detritus food webs. The relationship between feeding habits and radioactive Cs levels showed that earthworms, representing the detritivores, had radioactive Cs levels about 85 times higher than grasshoppers which rely on the grazing food web. We reported that the ^{137}Cs concentration in the earthworms remained stable from 2014 to 2016 [19]. These findings clearly indicate that the detritus food web is more highly contaminated than the grazing food web. High levels of ^{137}Cs accumulation flow up to higher trophic levels through the detritus food web [10]. Therefore, the high levels of radioactive Cs concentration maintained in Jorô spiders during this study period could be explained by their food resources coming from the contaminated detritus food web. In forests, the biomass of aerial insects from the detrital food web increases in spring and autumn, and web spider depends on these aerial insects [23]. In the present study, Jorô spiders were collected in autumn between September and October; thus the food resources of Jorô spiders were relatively dependent on the detritus food web during the collection period. This study thus demonstrates that the detritus food web is the primary pathway for long-term movement of radioactive Cs through the food web. Moreover, the detritus food web makes a large contribution to the transfer and circulation of radioactive Cs within the ecosystem. From a long-term perspective, the behavior of radioactive Cs through the food web, including these invertebrate species, is important for understanding the environmental behavior of radioactive Cs and accurate radiation risk assessment for non-human species.

4.5   Conclusions

A 5-year study in arthropods showed variations in the chronological change of radioactive Cs concentration levels among arthropods of different trophic levels. The radioactive Cs concentration in both herbivorous grasshoppers and omnivorous field crickets significantly reduced from 2012 to 2016. In contrast, the level in carnivorous Jorô spiders did not change significantly during the survey period. This variance is likely caused by the difference in food resource pathways between the grazing food web and the detritus food web (Fig. 4.5). Detritivorous earthworms showed the highest radioactive Cs concentration, and the comparison between trophic levels and radioactive Cs concentration clearly showed high contamination in the detritus food web. This study demonstrates that the detritus food web is the primary pathway for long-term radioactive Cs movement through the food web. Long-term monitoring of terrestrial invertebrates is thus necessary to understand the behavior of radioactive Cs in the ecosystem and to contribute to risk assessment for non-human species.

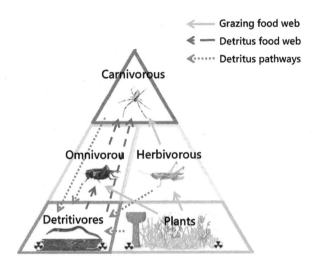

Fig. 4.5 Transfer pathways of radioactive Cs in terrestrial invertebrates. Each species represents feeding habits of terrestrial invertebrates; thus each species does not always prey on each other directly. For example, web spider cannot prey earthworm, but they can prey detritivores such as flies (Diptera) which emerge from the same habitat of earthworm. Therefore, earthworm represents radioactive Cs contamination in detritivores, and the other three species also represent the contamination of each feeding habit

References

1. Fukushima Prefectural Government (2015) Forest and Forestry Statistics. http://www.pref. fukushima.lg.jp/sec/36055a/toukeisyo.html. Accessed 1 Mar 2018
2. Steiner M, Linkov I, Yoshida S (2002) The role of fungi in the transfer and cycling of radionuclides in forest ecosystems. J Environ Radioact 58:217–241
3. Myttenaere C, Schell WR, Thiry Y, Sombre L, Ronneau C, van der Stegen de Schrieck J (1993) Modelling of Cs-137 cycling in forests: recent developments and research needed. Sci Total Environ 136:77–91
4. Wipfli MS (1997) Terrestrial invertebrates as salmonid prey and nitrogen sources in streams: contrasting old-growth and young-growth riparian forests in southeastern Alaska, U.S.A, Can J Fish Aquat Sci 54:1259–1269
5. Sato T, Watanabe K, Kanaiwa M, Niizuma Y, Harada Y, Lafferty KD (2011) Nematomorph parasites drive energy flow through a riparian ecosystem. Ecology 92:201–207
6. Nakano S, Miyasaka H, Kuhara N (1999) Terrestrial–aquatic linkages: riparian arthropod inputs alter trophic cascades in a stream food web. Ecology 80(7):2435–2441
7. Lavelle P, Bignell D, Lepage M, Wolters W, Roger P, Ineson P, Heal OW, Dhillion S (1997) Soil function in a changing world: the role of invertebrate ecosystem engineers. Eur J Soil Biol 33:159–193
8. Tyler AN, Carter S, Davidson DA, Long DJ, Tipping R (2001) The extent and significance of bioturbation on137Cs distributions in upland soils. Catena 43:81–99. 5
9. Ayabe Y, Kanasashi T, Hijii N, Takenaka C (2014) Radiocesium contamination of the web spider *Nephila clavata* (Nephilidae: Arachnida) 1.5 years after the Fukushima Dai-ichi Nuclear Power Plant accident. J Environ Radioact 127:105–110
10. Murakami M, Ohte N, Suzuki T, Ishii N, Igarashi Y, Tanoi K (2014) Biological proliferation of cesium-137 through the detrital food chain in a forest ecosystem in Japan. Sci Rep 4:1–5

11. Yoshimura M, Akama A (2013) Radioactive contamination of aquatic insects in a stream impacted by the Fukushima nuclear power plant accident. Hydrobiologia 722:19–30
12. Nakanishi H, Mori A, Takeda K, Tanaka H, Kobayashi N, Tanoi K, Yamakawa T, Mori S (2015) Discovery of radioactive silver (110mAg) in spiders and other fauna in the terrestrial environment after the meltdown of Fukushima Dai-ichi nuclear power plant. Proc Jpn Acad Ser B Phys Biol Sci 91:160–174
13. Ayabe Y, Kanasashi T, Hijii N, Takenaka C (2015) Relationship between radiocesium contamination and the contents of various elements in the web spider *Nephila clavata* (Nephilidae: Arachnida). J Environ Radioact 150:228–235
14. Sakai M, Gomi T, Negishi JN, Iwamoto A, Okada K (2016) Different cesium-137 transfers to forest and stream ecosystems. Environ Pollut 209:46–52
15. Tanaka S, Hatakeyama K, Takahashi S, Adati T (2016) Radioactive contamination of arthropods from different trophic levels in hilly and mountainous areas after the Fukushima Daiichi nuclear power plant accident. J Environ Radioact 164:104–112
16. Ishii Y, Hayashi S, Takamura N (2017) Radiocesium transfer in forest insect communities after the Fukushima Dai-ichi nuclear power plant accident. PLoS One 12:1–14
17. Hasegawa M, Ito MT, Kaneko S, Kiyono Y, Ikeda S, Makino S (2013) Radiocesium concentrations in epigeic earthworms at various distances from the Fukushima Nuclear Power Plant 6 months after the 2011 accident. J Environ Radioact 126:8–13
18. Hasegawa M, Kaneko S, Ikeda S, Akama A, Komatsu M, Ito MT (2015) Changes in radiocesium concentrations in epigeic earthworms inrelation to the organic layer 2.5 years after the 2011 Fukushima Dai-ichi nuclear power plant accident. J Environ Radioact 145:95–101
19. Tanaka S, Adati T, Takahashi T, Fujiwara K, Takahashi S (2018) Concentrations and biological half-life of radioactive cesium in epigeic earthworms after the Fukushima Dai-ichi Nuclear Power Plant accident. J Environ Radioact 192:227–232
20. R Core Team (2013) R: a language and environment for statistical computing. R Foundation for Statistical Computing, Vienna Austria ISBN 3-900051-07-0. http://www.R-project.org
21. IAEA (International Atomic Energy Agency) (2000) Generic procedures for assessment and response during a radiological emergency. Iaea-Tecdoc-1162.
22. Polis GA, Strong DR (1996) Food web complexity and community dynamics. Am Nat 147(5):813–846
23. Shimazaki A, Miyashita T (2005) Variable dependence on detrital and grazing food webs by generalist predators: aerial insects and web spiders. Ecography (Cop) 28:485–494

5

FNPP Accident and Strontium Pollution in the Environment

Kazuma Koarai, Yasushi Kino, Toshitaka Oka, Atsushi Takahashi,
Toshihiko Suzuki, Yoshinaka Shimizu, Mirei Chiba, Ken Osaka,
Keiichi Sasaki, Yusuke Urushihara, Tomokazu Fukuda, Emiko Isogai,
Hideaki Yamashiro, Manabu Fukumoto, Tsutomu Sekine,
and Hisashi Shinoda

Abstract Substantial amounts of radionuclides including strontium-90 (^{90}Sr) were released by the Fukushima Daiichi Nuclear Power Plant (FNPP) accident. In the present study, we describe and discuss the presence of ^{90}Sr in the ex-evacuation zone of the FNPP accident and its relationship with ^{90}Sr activity concentration in the hard tissue of animals. We found that the activity concentration of ^{90}Sr in the hard tissue exhibited a positive correlation with ^{90}Sr pollution in their corresponding terrestrial and marine environments. Hard tissues, such as the teeth, bones, and otoliths, of animals and fishes could serve as useful tools in assessing ^{90}Sr pollution in the environment during the period of the formation of those tissues.

K. Koarai (✉) · Y. Kino
Department of Chemistry, Tohoku University, Sendai, Japan
e-mail: koarai.kazuma@jaea.go.jp

T. Oka · T. Sekine (✉)
Department of Chemistry, Tohoku University, Sendai, Japan

Institute for Excellence in Higher Education, Tohoku University, Sendai, Japan
e-mail: tsekine@m.tohoku.ac.jp

A. Takahashi
Tohoku University Hospital, Tohoku University, Sendai, Japan

T. Suzuki · K. Osaka
Graduate School of Dentistry, Tohoku University, Sendai, Japan

International Research Institute of Disaster Science, Tohoku University, Sendai, Japan

Keywords [90]Sr · Hard tissue · Tooth · Bone · Cattle · Otolith · Fish · Fukushima Daiichi Nuclear Power Plant accident

5.1 Introduction

Strontium-90 (^{90}Sr) has been released to the environment from various nuclear disasters in the past, such as nuclear weapon tests [1], the Chernobyl Nuclear Power Plant accident [2, 3], release of contaminated radionuclides from nuclear fuel processing plants at Sellafield and Mayak facilities [4, 5], and sea disposal operations by the former Soviet Union [6]. Strontium-90 was released also by the Fukushima Daiichi Nuclear Power Plant (FNPP) accident in 2011 and remained in the terrestrial and marine environments due to its long physical half-life of 28.8 years. Its long biological half-life and bone-seeking property may have adverse effects on the bone marrow, and special attention should be paid to the behavior of [90]Sr. Strontium-90 easily transfers to terrestrial and marine biota because it is rather soluble [7, 8]. However, the relationship between ^{90}Sr in the environment and in animal body is still unclear. In this chapter, we describe and discuss the presence of ^{90}Sr in the environment and its migration into the teeth and bones of cattle abandoned in the ex-evacuation zone of the FNPP accident. We also discuss ^{90}Sr activity concentration in the otolith of marine fishes around FNPP. Through the discussion, we suggest that ^{90}Sr activity concentration in the hard tissue of animal body might reflect the extent of ^{90}Sr pollution in the environment.

Y. Shimizu · M. Chiba · K. Sasaki · H. Shinoda (✉)
Graduate School of Dentistry, Tohoku University, Sendai, Japan
e-mail: shinoda-h@m.tohoku.ac.jp

Y. Urushihara
Graduate School of Medicine, Tohoku University, Sendai, Japan

T. Fukuda
Faculty of Science and Engineering, Iwate University, Morioka, Japan

E. Isogai
Graduate School of Agricultural Sciences, Tohoku University, Sendai, Japan

H. Yamashiro
Faculty of Agriculture, Niigata University, Niigata, Japan

M. Fukumoto
Institute of Development, Aging and Cancer, Tohoku University, Sendai, Japan

School of Medicine, Tokyo Medical University, Tokyo, Japan

5.2 ^{90}Sr Pollution by the FNPP Accident

5.2.1 ^{90}Sr in the Environment

The Ministry of Education, Culture, Sports, Science and Technology, Japan (MEXT), reported that 0.1–6 kBq/m^2 of ^{90}Sr and 0.3–17 kBq/m^2 of ^{89}Sr were detected in soil within a 20-km radius from FNPP (the ex-evacuation zone) [9]. Strontium-90 is known to have been present in the Japanese environment before the FNPP accident, which probably stemmed from the atmospheric nuclear weapon tests conducted during the 1950s–1980s. However, the existence of ^{89}Sr in soil of the ex-evacuation zone indicates that the ^{90}Sr pollution was caused by the FNPP accident, because ^{89}Sr has a relatively short half-life of 50.5 days and more than 50 years have passed since the atmospheric nuclear weapon tests were discontinued.

It was reported that the amount of ^{90}Sr released by the FNPP accident into the atmosphere is smaller than those of volatile radionuclides (noble gases, iodine, tellurium, and cesium). The amount of released ^{90}Sr is estimated to be three orders of magnitude smaller than that of cesium-137 (^{137}Cs) [3]. The nuclear fuel of FNPP did not exceed the temperature of 2,700 K that is necessary for the volatilization of refractory elements, such as ^{90}Sr and actinides [10]. Schwantes et al. estimated that most of radioactive Sr retained inside the reactors [11]. These reports suggest that the area polluted by ^{90}Sr might be smaller than that by ^{137}Cs attributed to the FNPP accident.

It was reported that ^{90}Sr activity concentration in soil and vegetation samples is one to four orders of magnitude lower than that of ^{137}Cs [12]. Sahoo et al. showed that ^{90}Sr activity concentration in soil from the ex-evacuation zone of the FNPP accident ranged from 3 to 23 Bq/kg, whereas ^{137}Cs activity concentration ranged from 0.7 to 110 kBq/kg [13], which is consistent with the previous reports [3, 10]. The ratio of ^{90}Sr/^{137}Cs in soil widely varies depending on the area examined.

The FNPP accident caused ^{90}Sr pollution not only in the terrestrial environment but also in the marine environment. Since the most of ^{90}Sr was presumed to remain in the reactor of FNPP, atmospheric release of ^{90}Sr was speculated to be minor. The total amount of ^{90}Sr released into the sea was estimated in the range from 0.09–0.9 PBq [14] to 1–6.5 PBq [15], and ^{90}Sr activity concentration in surface seawater near the harbor of FNPP was 0.2–400 kBq/m^3 [15]. The activity concentration of ^{90}Sr was an order lower than that of ^{137}Cs from April 2011 to February 2012, except for December 2011 when ^{90}Sr and ^{137}Cs activity concentrations were equivalent due to the discharge of ^{90}Sr-contaminated wastewater [16]. In the Pacific Ocean 15 km east from FNPP, ^{90}Sr activity concentration was two orders lower than that in seawater near FNPP. Owing to diffusion in offshore regions of Fukushima Prefecture, the pollution is believed to be limited to the sea around FNPP. These suggest that the ^{90}Sr pollution of the terrestrial and marine environments are limited to the vicinity of FNPP.

5.2.2 ^{90}Sr Activity Concentration in the Hard Tissue of Abandoned Cattle After the FNPP Accident

We previously reported activity concentration and specific activity of ^{90}Sr in the teeth of cattle stayed in the ex-evacuation zone (Fig. 5.1) of the FNPP accident [17]. Because of chemical similarity between Sr and calcium (Ca), the teeth incorporate Sr during their formation period (calcification period) and retain it until they fall out or are worn down. As the amount of ^{90}Sr accumulated in the teeth is assumed to reflect the amount of ^{90}Sr incorporated into the body during the formation of the teeth, we hypothesized that the assessment of ^{90}Sr activity concentration in the teeth might provide useful information about the degree of ^{90}Sr contamination in the environment.

Figure 5.2 shows the relationship between ^{90}Sr activity concentration in hard tissues (teeth and mandibular bones) of cattle and in soil from the ex-evacuation zone (Okuma Town and Kawauchi Village) after the FNPP accident. Activity concentrations of ^{90}Sr in the teeth were 150–830 mBq/g Ca (ratio of ^{90}Sr radioactivity to Ca weight in the tooth) (average, 470 mBq/g Ca) for Okuma Town and 100–310 mBq/g

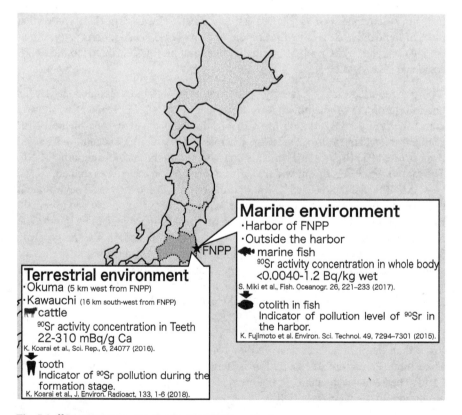

Fig. 5.1 ^{90}Sr pollution in animals after the FNPP accident

Ca (average, 170 mBq/g Ca) for Kawauchi Village [17]. The deposited amount of ^{90}Sr in soil was 94–1500 Bq/m^2 (average, 740 Bq/m^2) for Okuma Town and 39–380 Bq/m^2 (average, 200 Bq/m^2) for Kawauchi Village [9]. We selected Iwate Prefecture as the control area because it is approximately 250 km north from FNPP and is considered free from FNPP-related ^{90}Sr pollution. As shown in Fig. 5.2a, ^{90}Sr activity concentration in cattle teeth was significantly correlated with that in soil; the higher the ^{90}Sr activity concentration in soil, the higher the activity concentration in cattle teeth. A similar correlation was also found between ^{90}Sr activity concentration in soil and that in the mandibular bone (Fig. 5.2b). Metabolic turnover rates (modeling and remodeling) of the mandibular bone in the cattle are not well-

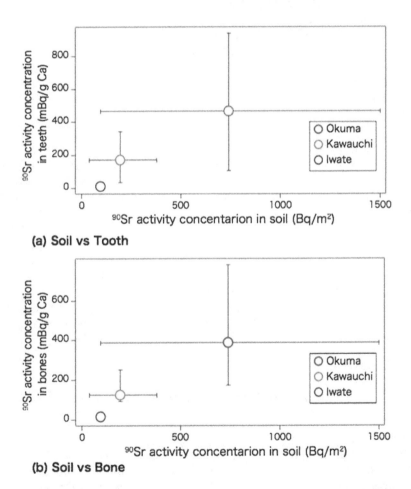

(a) Soil vs Tooth

(b) Soil vs Bone

Fig. 5.2 Relationship between ^{90}Sr activity concentration in hard tissues and that in soil. (**a**) Relationship between teeth and soils. (**b**) Relationship between bones and soils. Data of ^{90}Sr activity concentration in the teeth were obtained from our previous report [17]. Data of ^{90}Sr activity concentration in soil were obtained from Japanese government reports [9, 18]. Each symbol indicates the average value with error bars showing the maximum and minimum values

known. However, a major part of the original bone that formed before the FNNP accident was believed to be replaced by new bone formed after the accident, because, at the time the cattle were killed, 16–25 months had already passed since the accident. Thus, ^{90}Sr activity concentrations in hard tissues (both in the teeth and bones) reflect environmental ^{90}Sr levels during the formation of the tissues.

5.2.3   [90]Sr Pollution in Fish

Miki et al. reported ^{90}Sr activity concentrations in marine fish before and after the FNPP accident (Fig. 5.1) [19]. Activity concentrations higher than the background level before the FNPP accident were detected in 4 of 26 specimens collected just outside of the ex-evacuation zone, whereas ^{90}Sr activity concentrations in all samples collected offshore the neighborhood of Fukushima Prefecture were under the detection limit. The report pointed out that ^{90}Sr was detected only in marine fish living in the ocean near FNPP. Higher ^{137}Cs activity concentration was found in the whole body of teleost fish; the activity ratio of ^{137}Cs to ^{90}Sr was 5–190 times higher than that before the accident.

Regarding ^{90}Sr activity concentration in marine biota, the otolith of fish is a hard tissue composed of $CaCO_3$ and might be a unique index of ^{90}Sr pollution. In the process of otolith calcification, Sr is incorporated with Ca and is not metabolized once it is formed. It is reported that the β-ray count rate from otoliths of Japanese rockfish correlates with the activity concentration of radioactive Cs and ^{90}Sr in the whole body of fishes in the main harbor of FNPP. On the other hand, no β-rays were detected from fish collected outside of the main harbor of FNPP [20].

The otolith could reflect ^{90}Sr pollution in the living area of the fish. However, the use of otoliths as a biomonitor may not be feasible for detecting low levels of pollution, such as that outside FNPP harbor, because the amount of ^{90}Sr would be too small in the otolith. If more sensitive determination methods, such as mass-spectrometric quantification [21], were developed, ^{90}Sr in otoliths could be determined even with low radioactivity of the nuclide.

5.2.4   Migration of ^{90}Sr from the Environment to Hard Tissues of Animals

Strontium in soil exists in two forms; one is an insoluble solid form, and the other is the soluble form (exchangeable fraction) that can be dissolved in water or other aqueous solutions such as ammonium acetate [22, 23]. Considering the migration of ^{90}Sr from soil to hard tissues of animals, ^{90}Sr in the latter form is believed to be

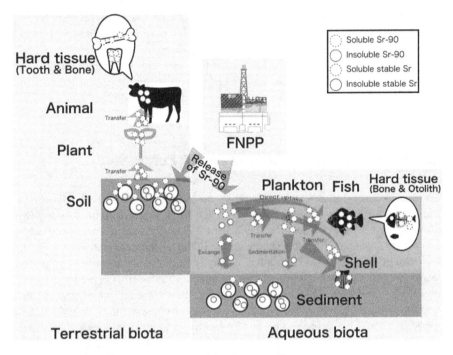

Fig. 5.3 Diagram of ^{90}Sr transfer in terrestrial and aqueous biota

important because Sr isotopes in the exchangeable fraction would be transferable in terrestrial biota. According to our analysis of soil from the ex-evacuation zone [23], only 1–9% of Sr in soil was in the exchangeable fraction, whereas 37–84% of ^{90}Sr in soil was in the exchangeable fraction. Consequently, the specific activity of ^{90}Sr (^{90}Sr/total Sr ≈ stable Sr) in the solid form was low (23–33 Bq/g Sr in Kawauchi Village and 11–36 Bq/g Sr in Okuma Town), and specific activity in the exchangeable fraction was high (200–330 Bq/g Sr in Kawauchi Village and 130–1200 Bq/g Sr in Okuma Town). As both ^{90}Sr and stable Sr behave similarly in the biota, the specific activity of ^{90}Sr would basically be identical in the migration route of ^{90}Sr from soil to plants and plants to animals (Fig. 5.3). In fact, specific activities of ^{90}Sr in cattle teeth (260–640 Bq/g Sr in Kawauchi Village and 380–1400 Bq/g Sr in Okuma Town) were in a similar level as those in the exchangeable fraction in soil.

After the FNPP accident, cattle in the ex-evacuation zone were released to the polluted field and allowed to graze on contaminated grasses. Once it enters the body, Sr is known to be incorporated in hard tissues during their formation period and accumulate there. The main chemical form of Sr in the teeth and bones is hydroxyapatite, $Ca_{10}(PO_4)_6(OH)_2$, in which any fraction of Ca may be replaced by Sr with little change in the apatite structure. ^{90}Sr incorporated in the hard tissue may exist as stable $Sr_XCa_{(10-X)}(PO_4)_6(OH)_2$ in the apatite crystal. Thus, ^{90}Sr activity concentration in the hard tissue of animals reflects the environmental pollution level during the formation of the tissue. Soluble form of Sr may migrate through the food chain and get directly incorporated into fish body along with the absorption of Ca^{2+}. ^{90}Sr would be accumulated in the hard tissues of fish, such as the bones, teeth, and otoliths. Therefore, the hard tissue of fishes may also provide useful information about ^{90}Sr pollution in the marine biota, as previously reported [20].

5.3   Summary and Perspectives for the Future Study

We described several aspects of ^{90}Sr pollution in the environment after the FNPP accident and ^{90}Sr activity concentration in the hard tissue of animal body. This article reviewed ^{90}Sr pollution in soil and seawater, hard tissues of abandoned cattle, and contaminated marine fishes around FNPP after the accident. Although the degree of environmental pollution has gradually been declining for the last several years, continuous research is necessary for clarifying ^{90}Sr mobility in the environment affected by the FNPP accident. Hard tissues such as the teeth, bones, and otoliths have the potential to reflect ^{90}Sr pollution of the environment during their formation period. It should be emphasized that ^{90}Sr specific activity in the teeth during their formation period and that in the exchangeable fraction in soil is at the same level [23]. This means that ^{90}Sr specific activity in the teeth may serve as a useful tool for understanding the migration route of ^{90}Sr in terrestrial biota. Furthermore, as the amount of ^{90}Sr in the teeth may be proportional to the amount of ^{90}Sr incorporated into the body during teeth formation, ^{90}Sr in the teeth provides a useful index for the individual assessment of internal exposure to radiation. Estimation of individual exposure dose is essential for understanding the biological effect of radiation, especially animals affected by the FNPP accident. In this sense, the examination of radioactive nuclides including ^{90}Sr in hard tissues provides useful information about the exposure dose to radiation. Nevertheless, further studies are required to elucidate the relationship between ^{90}Sr in hard tissues and internal exposure dose to the nuclide.

References

1. United Nations Scientific Committee on the Effects of Atomic Radiation (2000) Sources and effects of ionizing radiation, vol I. UNSCEAR 2000 Rep. pp 193–291
2. Higley KA (2006) Environmental consequences of the Chernobyl accident and their remediation: twenty years of experience. International Atomic Energy Agency, Vienna. ISBN, p 166

3. Steinhauser G, Brandl A, Johnson TE (2014) Comparison of the Chernobyl and Fukushima nuclear accidents: a review of the environmental impacts. Sci Total Environ 470–471:800–817. https://doi.org/10.1016/j.scitotenv.2013.10.029

4. Smith JN, Ellis KM, Boyd T (1999) Circulation features in the central Arctic Ocean revealed by nuclear fuel reprocessing tracers from Scientific Ice Expeditions 1995 and 1996. J Geophys Res 104:29663. https://doi.org/10.1029/1999JC900244

5. Jones S (2008) Windscale and Kyshtym: a double anniversary. J Environ Radioact 99:1–6. https://doi.org/10.1016/j.jenvrad.2007.10.002

6. Yablokov AV (2001) Radioactive waste disposal in seas adjacent to the territory of the Russian Federation. Mar Pollut Bull 43:8–18. https://doi.org/10.1016/S0025-326X(01)00073-X

7. International Atomic Energy Agency (2004) Sediment distribution coefficients and concentration factors for biota in the marine environment. Technical Reports Series No. 422

8. International Atomic Energy Agency (2010) Handbook of parameter values for the prediction of radionuclide transfer in terrestrial and freshwater. Technical Reports Series No. 472

9. Ministry of Education, Culture, Sports, Science and Technology, Japan (2011) Results of the nuclide analysis of Plutonium and Strontium by MEXT. Available at: http://radioactivity.nsr.go.jp/en/contents/5000/4167/24/1750_093014.pdf. Accessed 1 Sept 2015

10. Kirchner G, Bossew P, De CM (2012) Radioactivity from Fukushima Dai-ichi in air over Europe; part 2: what can it tell us about the accident? J Environ Radioact 114:35–40. https://doi.org/10.1016/j.jenvrad.2011.12.016

11. Schwantes JM, Orton CR, Clark RA (2012) Analysis of a nuclear accident: fission and activation product releases from the Fukushima Daiichi Nuclear Facility as remote indicators of source identification, extent of release, and state of damaged spent nuclear fuel. Environ Sci Technol 46:8621–8627. https://doi.org/10.1021/es300556m

12. Steinhauser G, Schauer V, Shozugawa K (2013) Concentration of strontium-90 at selected hot spots in Japan. PLoS One 8:1–5. https://doi.org/10.1371/journal.pone.0057760

13. Sahoo SK, Kavasi N, Sorimachi A et al (2016) Strontium-90 concentration in soil samples from the exclusion zone of the Fukushima Daiichi nuclear power plant. Sci Rep 6:23925. https://doi.org/10.1038/srep23925

14. Casacuberta N, Masqué P, Garcia-Orellana J et al (2013) ^{90}Sr and ^{89}Sr in seawater off Japan as a consequence of the Fukushima Dai-ichi nuclear accident. Biogeosciences 10:3649–3659. https://doi.org/10.5194/bg-10-3649-2013

15. Povinec PP, Gera M, Holý K et al (2013) Dispersion of Fukushima radionuclides in the global atmosphere and the ocean. Appl Radiat Isot 81:383–392. https://doi.org/10.1016/j.apradiso.2013.03.058

16. Tokyo Electric Power Co (2012) http://www.tepco.co.jp/cc/press/2012/1201877_1834.html. Accessed: 1th Sept 2018

17. Koarai K, Kino Y, Takahashi A et al (2016) ^{90}Sr in teeth of cattle abandoned in evacuation zone: record of pollution from the Fukushima-Daiichi Nuclear Power Plant accident. Sci Rep 6:24077. https://doi.org/10.1038/srep24077

18. Ministry of Education, Culture, Sports, Science and Technology, Japan (2009) 51th bulletin of research results of environmental radioactivity. Available at: http://www.kankyo-hoshano.go.jp/08/ers_lib/ers_abs51.pdf. Accessed: 1th Sept 2018

19. Miki S, Fujimoto K, Shigenobu Y et al (2017) Concentrations of ^{90}Sr and ^{137}Cs/^{90}Sr activity ratios in marine fishes after the Fukushima Dai-ichi Nuclear Power Plant accident. Fish Oceanogr 26:221–233. https://doi.org/10.1111/fog.12182

20. Fujimoto K, Miki S, Kaeriyama H et al (2015) Use of otolith for detecting strontium-90 in fish from the harbor of Fukushima Dai-ichi Nuclear Power Plant. Environ Sci Technol 49:7294–7301. https://doi.org/10.1021/es5051315

21. Takagai Y, Furukawa M, Kameo Y et al (2014) Sequential inductively coupled plasma quad-
    rupole mass-spectrometric quantification of radioactive strontium-90 incorporating cascade
    separation steps for radioactive contamination rapid survey. Anal Methods 6:355–362. https://
    doi.org/10.1039/C3AY41067F
22. Krouglov SV, Kurinov AD, Alexakhin RM (1998) Chemical fractionation of ^{90}Sr, ^{106}Ru, ^{137}Cs,
    and ^{144}Ce in Chernobyl-contaminated soils: an evolution in the course of time. J Environ
    Radioact 38:59–76. https://doi.org/10.1016/S0265-931X(97)00022-2
23. Koarai K, Kino Y, Takahashi A et al (2018) ^{90}Sr specific activity of teeth of abandoned cattle
    after the Fukushima accident – teeth as an indicator of environmental pollution. J Environ
    Radioact 183:1–6. https://doi.org/10.1016/j.jenvrad.2017.12.005

Part II
Consequences of Exposure to Radiation in Wild Organisms

6

Impact on Invertebrates in the Intertidal Zone

Toshihiro Horiguchi, Keita Kodama, Gen Kume, and Ik Joon Kang

Abstract In June 2014, May and June 2015, and June 2016, we conducted quantitative quadrat surveys of sessile invertebrates at seven intertidal sites in Ibaraki, Fukushima, and Miyagi Prefectures, including the sites near Fukushima Daiichi Nuclear Power Plant (FNPP), to check whether species richness, population densities, and biomass had recovered from declines after the 2011 Tohoku earthquake, tsunami and nuclear disaster. Additionally, in April, July, August and September from 2012 to 2017, we monitored the population density and spawning behavior of rock shells (*Thais clavigera*) in the field near FNPP. Increases in species richness and population densities in the intertidal zone near FNPP were not found until at least 4–5 years had passed after the FNPP accident. Densities of and reproductive performance by *T. clavigera* populations near FNPP in 2017 remained below levels before the accident. Although invertebrate larval recruitment from remote areas to the intertidal zone near FNPP could have been expected, this was not clearly observed until 2016 at the earliest. Thus, it is possible that environmental factors inhibited invertebrate reproduction, recruitment or both in the intertidal zone near FNPP at least for 5 years.

Keywords Fukushima Daiichi Nuclear Power Plant · Intertidal biota · Species richness · Population decline · Population density · Recovery · Recruitment · Reproduction · Rock shell (*Thais clavigera*) · Sessile organisms

T. Horiguchi (✉) · K. Kodama
National Institute for Environmental Studies, Tsukuba, Ibaraki, Japan
e-mail: thorigu@nies.go.jp

G. Kume
Kagoshima University, Kagoshima, Japan

I. J. Kang
Kyushu University, Fukuoka, Japan

6.1   Introduction

After the 2011 Tohoku earthquake (M_w 9.0) and tsunami on 11 March 2011(the Great East Japan Earthquake Disaster), three nuclear reactors melted down at Fukushima Daiichi Nuclear Power Plant (FNPP). Hydrogen explosions in reactor buildings resulted in the emission of hundreds of petabecquerels (PBq) of radionuclides to the environment [1]. The amount of radionuclide leakage from the FNPP accident was about one-tenth of that released by the 1986 Chernobyl NPP accident in Ukraine, where the total release of radionuclides was estimated to be 5,300 PBq, excluding radioactive noble gases (e.g., krypton-85, xenon-137 etc) [1].

The Tokyo Electric Power Company (FNPP owner and operator, TEPCO) estimated that 500 PBq of radioactive noble gases, 500 PBq of iodine-131 (^{131}I), 10 PBq of cesium-134 (^{134}Cs) and 10 PBq of ^{137}Cs were released from FNPP to the atmosphere between March 12 and 31, 2011 [2] and that atmospheric fallout and direct leakage from the reactors released an additional 11 PBq of ^{131}I, 3.5 PBq of ^{134}Cs and 3.6 PBq of ^{137}Cs into the marine environment between March 26 and September 30, 2011 [3]. Meanwhile, total deposition of ^{137}Cs from the atmosphere onto the ocean surface is estimated to have been 5–15 PBq [4–8]. Estimates of direct ^{137}Cs leakage from FNPP into the sea range from 3 to 6 PBq [5–13].

Four major sources released FNPP-derived radionuclides to the environment. In the order of decreasing magnitude, these were as follows:

1. The initial venting and explosive releases of gases and volatile radionuclides to the atmosphere, the largest and earliest source, which led to fallout on land and at sea.
2. Direct leakage of contaminated material, including radionuclides from the reactors to the sea during emergency cooling efforts at FNPP.
3. Ongoing radionuclide release via groundwater discharge.
4. Ongoing radionuclide release via river runoff.

Radionuclide releases via groundwater and river discharge were much smaller than the initial atmospheric fallout and the subsequent direct leakage [14].

Although transport models indicate that more than 80% of the atmospheric fallout derived from the FNPP accident would have been on the ocean surface, with the highest levels of deposition expected in coastal waters near FNPP. However, there are no atmospheric fallout data over the ocean to confirm this [14].

The severity of the FNPP accident raised concerns about contamination of aquatic organisms by radionuclides, in both freshwater and marine environments. By the end of March 2011, the Japanese government began to determine activity concentration of radionuclidess (i.e., γ-emitters) in aquatic organisms (i.e., fish and shellfish) to protect human consumers. The analytical results are available on the website of the Fisheries Agency of Japan [15]. In general, contamination of marine organisms by radioactive Cs is higher in demersal fish than in pelagic fish; radionuclide contamination levels in both demersal and pelagic fish collected off Fukushima Prefecture have been higher than those in similar fish collected off other prefectures [15]. The activity concentrations of radioactive Cs in fish tissue, however, have since decreased in most fish sampled from the region (e.g., Wada et al. [16]). Fishing operations in the Fukushima Prefecture region restarted on a trial basis in June 2012, and the areas and species targeted have gradually expanded since then [17].

Since the FNPP accident occurred, there have been many researches and activities in the marine environment, including coastal waters off Fukushima Prefecture, to analyze spatiotemporal changes in radionuclide activity concentration, to determine the final fate of the radionuclides introduced into seawater and to evaluate contamination levels in marine organisms. However, within 20 km of FNPP, there are few available data on the distribution and spatiotemporal changes of radionuclides emitted from FNPP and their possible ecological effects.

According to the literature, wildlife including invertebrates, is tolerant to varying degrees of γ-radiation; at 100–1000 mGy/day, some mortality can be expected in larvae and hatchlings of flatfish [18]. At 10–100 mGy/day, reduced reproductive success is observed in flatfish, and at 1–10 mGy/day, reduced reproductive success due to reduced fertility is possible [18]. Invertebrates such as crabs are more tolerant of radiation than are flatfish [18]. Estimated acute lethal dose (LD_{50}) is >100 Gy for marine invertebrates, 10–25 Gy for fishes, and 0.16 Gy for fish (salmonid) embryos [19]. Chronic exposure has yielded no-observed-adverse-effect dose rate of 10–30 mGy/h (= 240–720 mGy/day) for mortality and 3.2–17 mGy/h (= 76.8–408 mGy/day) for reproduction in snails, marine scallops, clams and crabs [19]. The no-observed-effect dose rate for fish reproduction is 1 mGy/h (= 24 mGy/day) [19]. The United Nations Scientific Committee on the Effects of Atomic Radiation (UNSCEAR) extensively analyzed the relevant data on the effect of radiation on the environment and on nonhuman biota [20, 21], concluding that maximum dose rate of less than 400 μGy/h (= 9.6 mGy/day) to any individual aquatic organism would be unlikely to have any detrimental effects at the population level [22]. This is based on the knowledge that there is little consistent and substantial evidence for any effect on reproduction at dose rate of <200 μGy/h (= 4.8 mGy/day) [23–25]. A generic dose rate of 10 μGy/h (= 240 μGy/day) is suggested for use in screening out environmental exposure situations of negligible concern [25–27].

On the other hand, anaseismic effects and the effect of tsunami backwash on the coastal organisms of eastern Japan have been estimated to be large or serious [28, 29]. The tsunami also caused the loss of tidal flats, as well as subsidence, temporary land elevation, and changes in sediment composition such as sludging or muddying, all of which had either direct or indirect impacts on the benthic organisms inhabiting the coastal areas [30]. The impacts of the tsunami on coastal benthic organisms, and the severity of these impacts, vary among species and sites [28, 29]. Temporary population declines have been observed in the Manila clam (*Ruditapes philippinarum*) in Matsukawaura, Fukushima Prefecture, although no effects on distribution have been observed in the bivalve *Gomphina melanegis* in Otsuchi Bay, Iwate Prefecture [31, 32]. The tsunami also affected reef resources, including abalone and sea urchins, through loss or reduction of seaweed forests as well as through direct impacts, though in Miyagi and Iwate Prefectures, the populations of some of these species have been observed to recover [32, 33].

Although dose rate from the environmental exposure in intertidal invertebrates near FNPP after the accident has not yet been estimated, Horiguchi et al. [34] investigated the intertidal zone from Shirahama (Chiba Prefecture) to Kuji (Iwate Prefecture) along the coastline of eastern Japan (a total distance of ca. 800 km) in 2011, 2012 and 2013 to evaluate the ecological effect of the FNPP accident that accompanied the Great East Japan Earthquake Disaster. They observed that the number of intertidal species decreased significantly with proximity to FNPP and that no rock shell (*Thais clavigera*) specimens were collected near FNPP, from Hirono to Futaba Beach (a distance of approximately 30 km) in 2012. The collection of rock shell specimens at many other sites hit by the tsunami suggested that the absence of rock shells around FNPP in 2012 might have been caused by the FNPP accident in 2011. Quantitative surveys in 2013 showed that the number of species and population densities in the intertidal zones were much lower at sites near or within several kilometers south of FNPP than at other sites and lower than in 1995, especially in the case of Arthropoda. There is no clear explanation for these findings, but evidently, the intertidal biota around FNPP has been harmed since the nuclear accident [34].

In June 2014, May and June 2015, and June 2016, we conducted quantitative quadrat surveys of sessile invertebrates at seven intertidal sites previously surveyed in Ibaraki, Fukushima and Miyagi Prefectures including sites near FNPP, to evaluate whether the intertidal invertebrates had recovered from the declines that followed the accident; we measured population densities (i.e., abundance, as indicated by the number of individuals per square meter), biomass (total wet weight of invertebrates per square meter) and species numbers. Additionally, in April, July, August, and September from 2012 to 2017, we monitored the density of rock shell (*T. clavigera*) populations and counted their egg capsules in the field near FNPP.

6.2   Materials and Methods

On June 13-16 and 26-28, 2014, May 18-21 and June 17-19, 2015, June 4-7, 21, 23 and 24, 2016, quantitative quadrat surveys of sessile invertebrates were conducted at seven sites: Tomioka Fishing Port (Tomioka Town; shown as Tomioka in Figs. 6.1b, 6.2, 6.3, 6.4, 6.5, 6.6 and 6.7), Ottozawa coast (Okuma Town; shown as Okuma in Figs. 6.1b, 6.2, 6.3, 6.4, 6.5, 6.6 and 6.7), Kubo-yaji coast (Futaba Town; shown as Kuboyaji in Figs. 6.1b, 6.2, 6.3, 6.4, 6.5, 6.6 and 6.7) and Urajiri coast (Minami-Soma City; shown as Urajiri in Figs. 6.1b, 6.2, 6.3, 6.4, 6.5, 6.6 and 6.7) in Fukushima Prefecture within a 20-km radius of FNPP; Hasaki Beach (Kamisu City) and Kujihama Beach (Hitachi City) in Ibaraki Prefecture; and Watanoha coast (Ishinomaki City; shown as Ishinomaki in Figs. 6.1b, 6.2, 6.3, 6.4, 6.5, 6.6 and 6.7) in Miyagi Prefecture as reference sites for comparison (Fig. 6.1b and Table 6.1). These sites were the same as those surveyed in 2013 and were representative of

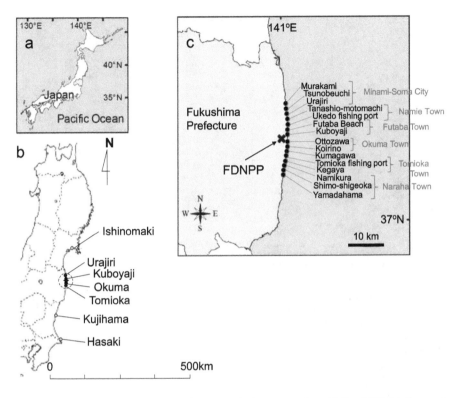

Fig. 6.1 Sampling sites of intertidal invertebrates during surveys from 2012 to 2017. (**a**) A map of Japan, where Fukushima Prefecture is marked in yellow. (**b**) Locations for sampling intertidal invertebrates during surveys in 2014, 2015 and 2016. Purple star marks the location of Fukushima Daiichi Nuclear Power Plant (FNPP). Dashed circle on map indicates a radius of 20 km from FNPP. Red dots on map indicate sites located within the 20-km radius of FNPP. (**c**) Site locations for surveys on density and spawning behavior of *Thais clavigera* populations from 2012 to 2017. FNPP is located between Ottozawa and Kubo-yaji

those used in the 2012 surveys in terms of substrate (i.e., tetrapods or similar concrete structures set along the coast for wave protection) as well as distance from FNPP [34]. Sessile organisms on the surface of tetrapods or similar concrete structures within a 50 × 50-cm quadrat were collected at three different elevations in the intertidal zone (the lower, middle and upper intertidal zones) at each site. Specimens were preserved in neutral buffered 10% formalin solution. After identifying the species, the number of individuals and wet weight were determined for each species on the spot, along with the sampling location and elevation. Additionally, we conducted field observations on the density and spawning behavior of *T. clavigera* populations near FNPP in April, July, August and September from 2012 to 2017 (Fig. 6.1c and Table 6.2). The density of *T. clavigera* populations was expressed by the number of individuals collected per minute, because it was difficult to quantitatively represent

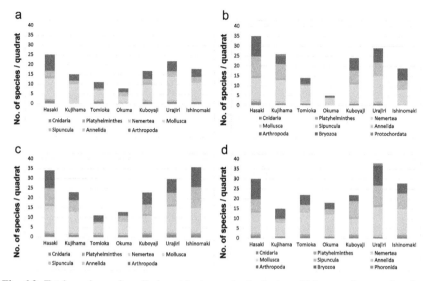

Fig. 6.2 Total numbers of sessile invertebrate species in the intertidal zone of seven sites from 2013 to 2016, as sampled with a 50 × 50-cm quadrat. (**a**) 2013, (**b**) 2014, (**c**) 2015 and (**d**) 2016. FNPP is located between Okuma and Kubo-yaji. Tomioka, Okuma, Kubo-yaji and Urajiri are located within the 20-km radius of FNPP. Distances between sites do not correspond to bar locations on the charts

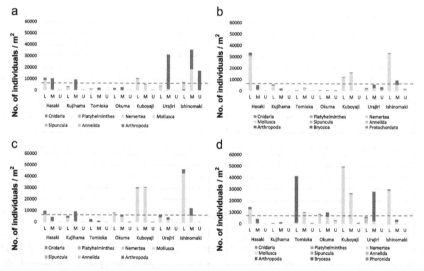

Fig. 6.3 Densities of sessile invertebrates (number/m²) in the intertidal zone (by elevation) from 2013 to 2016. (**a**) 2013, (**b**) 2014, (**c**) 2015 and (**d**) 2016. *L* lower intertidal zone, *M* middle intertidal zone, *U* upper intertidal zone. Data were collected with a 50 × 50-cm quadrat. FNPP is located between Okuma and Kubo-yaji. Tomioka, Okuma, Kubo-yaji and Urajiri are located within the 20-km radius of FNPP. Distances between sites do not correspond to bar locations on the charts. Pink dotted line represents the average number of individuals/m² from quadrat surveys of sessile organisms conducted in May 1995 at 20 sites along the coast of Fukushima Prefecture [37]: the average population density in 1995 was 7,158 individuals/m², consisting of Arthropoda (4593, 64.2%), Annelida (179, 2.5%), Mollusca (2348, 32.8%) and other organisms (38, 0.5%)

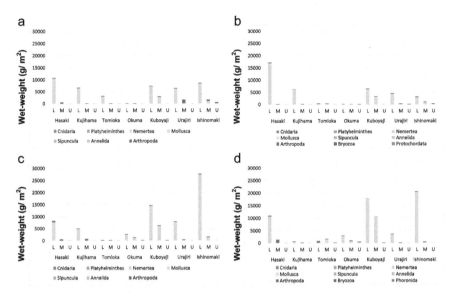

Fig. 6.4 Total biomass (wet weight, g/m²) of sessile invertebrates in the intertidal zone (by elevation) from 2013 to 2016. (**a**) 2013, (**b**) 2014, (**c**) 2015 and (**d**) 2016. *L* lower intertidal zone, *M* middle intertidal zone, *U* upper intertidal zone. Data were collected with a 50 × 50-cm quadrat. FNPP is located between Okuma and Kubo-yaji. Tomioka, Okuma, Kubo-yaji and Urajiri are located within the 20-km radius of FNPP. Distances between sites do not correspond to bar locations on the charts

the area where *T. clavigera* specimens were collected at different forms of tetrapods or concrete structures for wave protection.

To assess similarity in the biotic community structure of the intertidal zone in the survey from 2014 to 2016, we conducted cluster analysis (group average method) on Bray-Curtis similarity matrices for the number of species, population density (the number of individuals/m²) and total sessile biomass (whole wet weight per square meter, g/m²). Population density data from the lower, middle, and upper intertidal zones were merged within each site, and log $(1 + x)$ transformed for calculation of Bray-Curtis similarity. Site grouping was performed with a cut-off similarity level of 70 or 80%. Differences in the biotic community structure represented by Bray-Curtis similarity among site groups [35] were tested by analysis of similarity (ANOSIM). Possible differences in population densities of Arthropoda (i.e., the number of individuals/m²) at sites near FNPP (i.e., Tomioka, Okuma, and Kubo-yaji) compared to those at other sites (i.e., Hasaki, Kujihama, Urajiri, and Ishinomaki) were evaluated using a t-test under the assumption that they had approximately equal variances each year. Statistical analyses (t-test) were performed with Microsoft Excel 2016 statistical software, except for cluster analysis and ANOSIM, which were conducted with PRIMER6 software [36].

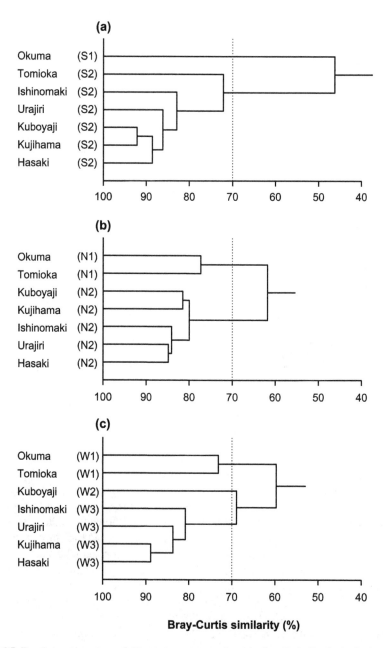

Bray-Curtis similarity (%)

Fig. 6.5 Dendrograms constructed by the group average method on Bray-Curtis similarity matrices for (**a**) number of species, (**b**) population density, and (**c**) total biomass of intertidal invertebrates collected in northeastern Japan on June 13–16 and 26–28, 2014. Population density data from the lower, middle, and upper intertidal zones were merged within each site, and log (1+x) transformed for calculation of Bray-Curtis similarity. Site groupings based on cut-off similarity level of 70% (dotted lines) are shown as S1-S2, N1-N2, and W1-W3 for panels (**a**), (**b**), and (**c**), respectively. The two sites (Okuma and Tomioka) located south of FNPP are shown in red text

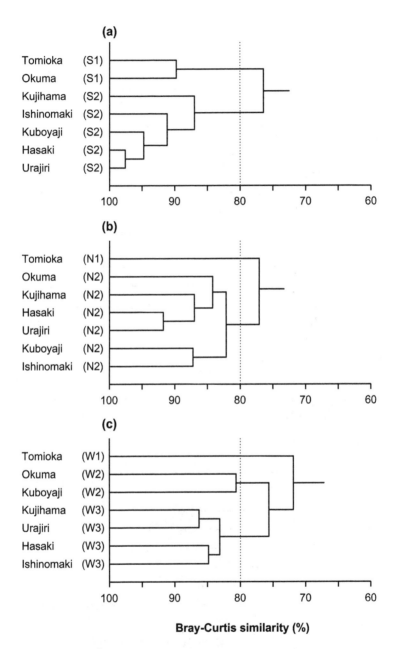

Fig. 6.6 Dendrograms constructed by the group average method on Bray-Curtis similarity matrices for (**a**) number of species, (**b**) population density, and (**c**) total biomass of intertidal invertebrates collected in northeastern Japan on May 18–21 and June 17–19, 2015. Population density data from lower, middle, and upper intertidal zones were merged within each site, and log (1+x) transformed for calculation of Bray-Curtis similarity. Site groupings based on cut-off similarity level of 80% (dotted lines) are shown as S1-S2, N1-N2, and W1-W3 for panels (**a**), (**b**), and (**c**), respectively. The two sites (Okuma and Tomioka) located south of FNPP are shown in red text

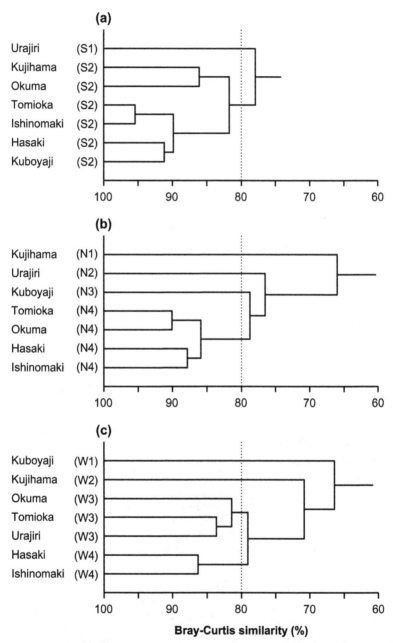

Fig. 6.7 Dendrograms constructed by the group average method on Bray-Curtis similarity matrices for (**a**) number of species, (**b**) population density, and (**c**) total biomass of intertidal invertebrates collected in northeastern Japan on June 4–7, 21, 23 and 24, 2016. Population density data from the lower, middle, and upper intertidal zones were merged within each site, and log (1+x) transformed for calculation of Bray-Curtis similarity. Site groupings based on cut-off similarity level of 80% (dotted lines) are shown as S1-S2, N1-N4, and W1-W4 for panels (**a**), (**b**), and (**c**), respectively

Table 6.1 Locations for sampling intertidal invertebrates during surveys in 2014, 2015 and 2016

2016 Survey Date	2015 Survey Date	2014 Survey Date	Sampling location				
			Prefecture	City/Town	Name of site	Latitude	Longitude
21-Jun	19-Jun	28-Jun	Ibaraki	Kamisu City	Hasaki Beach	35°45'14.8"N	140°50'16.3"E
23-Jun	17-Jun	26-Jun	Ibaraki	Hitachi City	Kujihama Beach	36°30'09.8"N	140°37'48.0"E
6-Jun	20-May	14-Jun	Fukushima	Tomioka Town	Tomioka Fishing Port	37°20'16.3"N	141°01'45.3"E
4-Jun	19-May	13-Jun	Fukushima	Okuma Town	Ottozawa coast	37°24'21.2"N	141°02'00.4"E
5-Jun	18-May	15-Jun	Fukushima	Futaba Town	Kubo-yaji coast	37°26'41.4"N	141°02'10.9"E
7-Jun	21-May	16-Jun	Fukushima	Minami-Soma City	Urajiri coast	37°31'33.7"N	141°01'53.8"E
		27-Jun	Miyagi	Ishinomaki City	Watanoha coast	38°24'50.8"N	141°21'28.1"E
	18-Jun		Miyagi	Ishinomaki City	Watanoha coast	38°24'48.4"N	141°21'31.6"E
24-Jun			Miyagi	Ishinomaki City	Watanoha coast	38°24'47.4"N	141°21'35.0"E

Fukushima Daiichi Nuclear Power Plant (FNPP) is located between Ottozawa and Kubo-yaji. Red text indicates sites within a 20-km radius of FNPP. We had to change the location at Watanoha, Ishinomaki a little every year, because of the restoration work of coastal structures

6.3 Results

In the intertidal zone at all sites surveyed from 2014 to 2016, the sessile species composition was dominated by the Mollusca and Arthropoda, followed by Annelida, similar to the 2013 results [34] (Fig. 6.2). The maximum number of intertidal species in 2014 was 35 at Hasaki, followed by 29 at Urajiri. The minimum of five species was at Okuma, located approximately 1 km south of FNPP (Fig. 6.2b). However, the similarity in species number between Tomioka and Okuma, both of which are located south of FNPP, was not very high and did not significantly differ from that between other sites in 2014, as determined by analysis of similarity (ANOSIM) of the biotic community structure represented by Bray-Curtis similarity among site groups [35] ($P = 0.14$; Figs. 6.2b and 6.5a). This result differed from that of the 2013 survey. The pattern of more species being identified in 2014 at Hasaki, Kujihama, Kubo-yaji, Urajiri and Ishinomaki than at Tomioka and Okuma was also observed in 2015 (Fig. 6.2c). The similarity in species number between Tomioka and Okuma, south of FNPP, was high and differed significantly from that between other sites in 2015 ($P < 0.05$; Figs. 6.2c and 6.6a). In 2016, however, no such pattern of more species being identified at Hasaki, Kujihama, Kubo-yaji, Urajiri and Ishinomaki than at Tomioka and Okuma was observed, due to an increase in the number of species at Tomioka and Okuma (Fig. 6.2d). The similarity in species number between Tomioka and Okuma, located south of FNPP, was not very high and differed insignificantly from that between other sites in 2016 ($P = 0.29$; Figs. 6.2d and 6.7a).

Population densities were higher in the lower and middle intertidal zones than in the upper intertidal zone at all sites surveyed from 2014 to 2016 (Fig. 6.3b–d), which is similar to the 2013 results [34]. Mollusca (e.g., mussels such as *Mytilus galloprovincialis* and *Septifer virgatus*) and Arthropoda (e.g., barnacles such as

Table 6.2 Site locations for surveys on density and spawning behavior of *Thais clavigera* populations from 2012 to 2017

2017 Survey Date	2016 Survey Date	2015 Survey Date	2014 Survey Date	2013 Survey Date	2012 Survey Date	Sampling Location Town/City	Name of site	Latitude	Longitude
26-Apr	8-Apr	19-Apr	15-Apr		24-Apr	Naraha	Yamadahama	37°15′29.1″N	141°00′50.0″E
26-Apr	8-Apr	19-Apr	15-Apr			Naraha	Shimo-shigeoka	37°16′39.1″N	141°01′04.6″E
26-Apr	8-Apr	19-Apr	15-Apr		24-Apr	Naraha	Namikura	37°18′25.2″N	141°01′28.1″E
27-Apr	11-Apr, 19-Jul	21-Apr, 30-Jul, 31-Aug	16-Apr, 13-Jul, 11-Aug	9-Aug	24-Apr	Tomioka	Kegaya	37°19′40.5″N	141°01′35.1″E
27-Apr, 25-Jul, 22-Aug	11-Apr, 19-Jul, 20-Aug	21-Apr, 30-Jul, 31-Aug	16-Apr, 13-Jul, 11-Aug, 10-Sep	9-Aug	24-Apr	Tomioka	Tomioka Fishing Port	37°20′16.3″N	141°01′45.3″E
			16-Apr, 13-Jul			Okuma	Kumagawa	37°23′16.7″N	141°02′03.0″E
28-Apr, 22-Jul	9-Apr, 20-Jul, 18-Aug	20-Apr, 31-Jul	13-Jul, 11-Aug, 10-Sep	9-Aug		Okuma	Koirino	37°23′26.5″N	141°02′04.5″E
28-Apr, 25-Jul, 22-Aug	9-Apr, 20-Jul, 18-Aug	20-Apr, 31-Jul, 31-Aug	17-Apr, 13-Jul, 11-Aug, 10-Sep	9-Aug	24-Apr	Okuma	Ottozawa	37°24′21.2″N	141°02′00.4″E
29-Apr, 26-Jul, 23-Aug	10-Apr, 21-Jul, 19-Aug	22-Apr, 1-Aug, 1-Sep	17-Apr, 14-Jul, 12-Aug, 11-Sep	10-Aug	25-Apr	Futaba	Kuboyaji	37°26′41.4″N	141°02′10.9″E

						Municipality	Site	Latitude	Longitude
29-Apr, 23-Aug	10-Apr, 21-Jul, 19-Aug	22-Apr, 1-Aug, 1-Sep	17-Apr, 14-Jul, 12-Aug, 11-Sep	10-Aug	25-Apr	Futaba	Futaba beach	37°27′12.1″N	141°02′20.7″E
29-Apr			18-Apr, 14-Jul		25-Apr	Namie	Ukedo fishing port	37°28′51.7″N	141°02′35.6″E
	10-Apr	23-Apr	18-Apr			Namie	Tanashio-motomachi	37°29′45.7″N	141°02′19.8″E
30-Apr, 26-Jul	12-Apr, 21-Jul, 19-Aug	23-Apr, 2-Aug, 2-Sep	19-Apr, 14-Jul, 12-Aug, 11-Sep	10-Aug	25-Apr	Minami-Soma	Urajiri	37°31′33.7″N	141°01′53.8″E
30-Apr	12-Apr		18-Apr		25-Apr	Minami-Soma	Tsunobeuchi	37°32′39.3″N	141°01′43.2″E
			19-Apr			Minami-Soma	Murakami	37°34′22.4″N	141°01′34.4″E

FNPP is located between Ottozawa and Kuboyaji. All sites were located within a 20-km radius of FNPP

Chthamalus challengeri) predominated at almost all sites surveyed. Species composition in 2014 was highly similar between Tomioka and Okuma, both located south of FNPP ($P < 0.05$; Figs. 6.3b and 6.5b), but not in 2015 ($P = 0.14$; Figs. 6.3c and 6.6b). In 2016, the species composition at Kujihama, Urajiri and Kubo-yaji differed significantly from that at the other sites (Tomioka, Okuma, Hasaki and Ishinomaki) ($P < 0.05$; Figs. 6.3d and 6.7b), but not from those at sites south of FNPP (i.e., Tomioka and Okuma), although it was one of typical characteristics in the 2013 survey that population densities in the intertidal zone were significantly lower at sites south of FNPP than at other sites [34].

From 2014 to 2016, the population density of sessile organisms at Kubo-yaji, approximately 1 km north of FNPP, was greater than those at Kujihama and Urajiri (Fig. 6.3b–d). However, the species composition of sessile organisms at Kubo-yaji differed markedly from those at other sites surveyed in that the Arthropoda accounted for less than 1% of all sessile organisms collected (Fig. 6.3b–d), which is similar to the 2013 survey [34]. Population densities of Arthropoda at sites near FNPP (Tomioka, Okuma, and Kubo-yaji) were significantly lower than those at other sites in 2014 and 2015 (*t*-test; $P < 0.05$), but not in 2016 ($P = 0.66$).

The combined wet weight of all sessile organisms from 2014 to 2016 was the greatest in the lower intertidal zone at almost all sites, followed by the middle and upper intertidal zone (Fig. 6.4b–d), similar to the results in 2013 [34]. From 2014 to 2016, molluscan biomass predominated at all surveyed sites, followed by arthropodan biomass (Fig. 6.4b–d). Species composition at Tomioka was similar to that at Okuma and differed significantly from that at other sites in 2014 and 2015 ($P < 0.05$; Figs. 6.4b, c, 6.5c, and 6.6c). In 2016, however, there was a significant difference of similarity in the species composition between the occurrences at Kubo-yaji, Kujihama, most central Fukushima sites (Okuma, Tomioka, and Urajiri), and the sites at both ends of the survey (Hasaki and Ishinomaki) ($P < 0.05$; Figs. 6.4d and 6.7c), but not in the south of FNPP (i.e., Tomioka and Okuma), which was one of the typical characteristics observed in the 2013 survey: total sessile biomass in the intertidal zone were significantly lower at sites south of FNPP than at other sites in the 2013 survey [34].

The distribution of rock shells (*T. clavigera*) near FNPP gradually expanded from sites in north (Minami-Soma City and Namie Town Fig. 6.8a, b, respectively) and south (Naraha Town and Tomioka Town Fig. 6.8f, e, respectively) to sites in central Fukushima (Futaba Town and Okuma Town Fig. 6.8c, d, respectively). Densities also seemed to increase from 2012 to 2016, with large variations (Fig. 6.8). When we started the field survey in December 2011, 9 months after the FNPP accident, *T. clavigera* had disappeared from Ottozawa, Okuma Town, approximately 1 km south of FNPP; they remained absent until July 2016: the first record on the density of *T. clavigera* population at Ottozawa, Okuma Town, in July 2016 was 0.03 individuals collected per minute (i.e., 3 individuals collected within 90 min). Consequently, the gap in the distribution of *T. clavigera* along the Fukushima coast seems to have closed by July 2016. However, population densities have remained low (around or less than 1 individual collected per minute) at

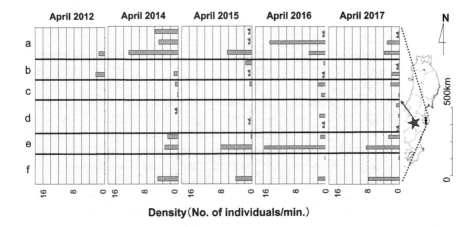

Fig. 6.8 Population densities (number/minute) of the rock shell (*Thais clavigera*) at 15 sites along the coast of Fukushima Prefecture from April 2012 to April 2017. Purple star marks the location of FNPP. Dashed circle on map indicates a radius of 20 km from FNPP. Red dots on map indicate sites located within the 20-km radius of FNPP. (**a**) Minami-Soma City, (**b**) Namie Town, (**c**) Futaba Town, (**d**) Okuma Town, (**e**) Tomioka Town, and (**f**) Naraha Town. *n.d.* no data

Table 6.3 Egg capsules in *Thais clavigera* populations at sites near FNPP

Location	Distance from FDNPP(km)	2013	2014	2015	2016	2017
Minami-Soma A (Urajiri)	17.5	++	+++	–	+++	+++
Futaba B (Beach)	3.5	–	–	++	–	+++
Futaba A (Kubo-yaji)	1.5	–	–	–	–	+
Okuma (Ottozawa)	1.2	–	–	–	–	++
Okuma (Koirino)	3.6	–	–	–	–	++
Tomioka B (Fishing port)	9.5	–	–	–	–	–
Tomioka A (Kegaya)	10.5	–	+	–	n.s.	n.s.

+++: > 100 cm² of the total areas where egg capsules were spawned
++: 5–100 cm² of the total areas where egg capsules were spawned
+: < 5 cm² of the total areas where egg capsules were spawned
–: no egg capsules observed
n.s. not surveyed
The names of locations are expressed as in [34]

sites in Okuma Town, within a few kilometers south of FNPP (Fig. 6.8d). Moreover, neither spawning behavior nor egg capsules were observed in populations at sites near FNPP (from Tomioka Fishing Port to Kubo-yaji) through the summer of 2016; egg capsules were first observed at Koirino, Ottozawa (Okuma Town) and Kubo-yaji (Futaba Town) in the summer of 2017 (Table 6.3 and Fig. 6.1c).

6.4   Discussion

In 2011, 2012 and 2013, Horiguchi et al. [34] investigated the ecological effect in the intertidal zone of eastern Japan affected by the FNPP accident that accompanied the Great East Japan Earthquake Disaster. They reported that the number of intertidal species decreased significantly with proximity to FNPP, and that no rock shell (*T. clavigera*) specimens were found near FNPP, from Hirono to Futaba Beach (a distance of approximately 30 km) in 2012. The presence of rock shell specimens at many other sites hit by the tsunami suggested that the absence of rock shells around FNPP in 2012 might have been caused by the FNPP accident. The quantitative 2013 surveys also showed that the number of species and population densities in the intertidal zones were much lower at sites near or within several kilometers south of FNPP than at other sites, and lower than in 1995, especially in the case of Arthropoda. There is no clear explanation for these findings. Although there are no estimates of radiation doses from environmental exposure in invertebrates in the intertidal zone near FNPP after the accident, clearly the intertidal biota around FNPP has been harmed since the nuclear accident.

The present study was carried out to investigate temporal changes in species composition and population densities in the intertidal zone of eastern Japan since 2014. Increases in species richness were evident at Tomioka Fishing Port (Tomioka) and Ottozawa (Okuma), which are located south of FNPP, from 2016, though this increase might have started at Ottozawa (Okuma) in 2015 (Fig. 6.2). Meanwhile, population densities have increased in the lower and middle intertidal zones at Ottozawa (Okuma) and Tomioka Fishing Port (Tomioka) since 2015 and 2016, respectively (Fig. 6.3). These increases were due to increases in small mussels (i.e., *Mytilus galloprovincialis*) at Ottozawa (Okuma) and barnacles (i.e., *Semibalanus cariosus* and *Chthamalus challengeri*) at Tomioka Fishing Port (Tomioka). In 1995, TEPCO conducted similar seasonal surveys using 30×30-cm quadrats at 20 sites in intertidal zone along the coast of Fukushima Prefecture, but only published a summary of their results [37]. In May 1995, they found an average of 7,158 individual sessile organisms/m^2, consisting of Arthropoda (4,593, 64.2%), Annelida (179, 2.5%), Mollusca (2348, 32.8%), and other organisms (38, 0.5%) [37]; thus, although Arthropoda clearly predominated, many other invertebrates were also present in Fukushima Prefecture in 1995, before the FNPP accident. Compared with the TEPCO data, we observed similar or higher numbers of individuals/m^2 in the lower and middle intertidal zones at Ottozawa (Okuma) in 2015 and Tomioka Fishing Port (Tomioka) in 2016 (Fig. 6.3). Because of their small sizes, however, increases in the total biomass at Ottozawa (Okuma) in 2015 and Tomioka Fishing Port (Tomioka) in 2016 were unclear (Fig. 6.4).

The species composition of sessile organisms at Kubo-yaji, approximately 1 km north of FNPP, differed markedly from that at other sites surveyed, in that the Arthropoda accounted for less than 1% of all sessile organisms collected (Fig. 6.3b–d). Although Arthropoda predominated in the intertidal zone of Fukushima Prefecture in 1995 [37], population densities seem to have decreased after the FNPP accident. The reduced population densities there have continued since 2013. Population densities of Arthropoda at sites near FNPP (Tomioka, Okuma, and Kubo-yaji) were significantly lower than those at other sites in 2014 and 2015 (t-test; $P < 0.05$), but not in 2016 ($P = 0.66$), possibly due to increases in barnacle population densities in the lower intertidal zone at Tomioka Fishing Port (Tomioka) in 2016 (Fig. 6.3d).

No *T. clavigera* specimens were observed near FNPP from Hirono to Futaba Beach (a distance of approximately 30 km) in 2012 [34]. In the present study conducted during 2014–2017, however, their distribution gradually expanded year by year from sites in the north and south to sites along the central Fukushima coast (Fig. 6.8), and the gap in the distributions, where no *T. clavigera* specimens were found, had closed by July 2016, approximately a year after an observation of increases in small mussels (i.e., *Mytilus galloprovincialis*) at Ottozawa (Okuma) in 2015; these mussels are prey for *T. clavigera* (Fig. 6.3c). The increase in prey organisms, such as small mussels (i.e., *Mytilus galloprovincialis*) at Ottozawa (Okuma) in 2015 might have been associated with the first observation of *T. clavigera* specimens there in July 2016, since our field survey started in December 2011, 9 months after the FNPP accident. The densities of *T. clavigera* populations seemed to increase from 2012 to 2016, with large variations (Fig. 6.8). Low densities in *T. clavigera* populations (around or less than 1 individual collected per minute), however, have persisted at sites in Okuma Town, within a few kilometers south of FNPP (Fig. 6.8d). In general, population densities of *T. clavigera* in Japan range from 1 to 4 individuals collected per minute (Horiguchi et al., unpublished data). Neither spawning behavior nor egg capsules spawned by *T. clavigera* populations were observed at sites near FNPP (from Tomioka Fishing Port to Kubo-yaji) through the summer of 2016 (Table 6.3). Egg capsules were first observed at Koirino, Ottozawa (Okuma Town), and Kubo-yaji (Futaba Town) in the summer of 2017, though the number of egg capsules spawned was limited (Table 6.3).

The number of species and population densities in the intertidal zone near FNPP did not begin to increase until at least 4–5 years after the FNPP accident. Even in 2017, 6 years after the disaster, densities of and reproductive performance by *T. clavigera* populations near FNPP remained below pre-disaster levels. Most invertebrates in the intertidal zone, including sessile organisms, have planktonic stages in their early life histories for certain periods: for example, *T. clavigera* has a veliger larva that lives a few months before settling [38]. In the coastal waters off Fukushima Prefecture, the generally weak currents and tides [39] are expected to effect the dispersion of larvae of most invertebrates in the intertidal zone, as well as dispersion of FNPP-derived radioactive Cs. The Kuroshio and Oyashio currents, which are the wind-driven boundary currents of the North Pacific, meet in the Kuroshio–Oyashio transition area off FNPP [40]. The Kuroshio is part of the subtropical gyre and transports warm saline water along the south coast of Japan and then eastward via the Kuroshio Extension (KE), whereas the Oyashio is part of the subarctic gyre and transports cold less saline water southward [41]. Actually, results on the bottom-sediment survey collected from a coastal strip (~30 × 120 km) off FNPP in October 2012 revealed that radioactive Cs concentration in the uppermost sediment layer were higher south of FNPP, though high activity concentration patches were also observed at a few sites north of FNPP [42]. Therefore, recruitment of their larvae could have been expected from remote areas to sites near FNPP. Nevertheless, data in the present study revealed that neither juvenile nor adult *T. clavigera* were observed at Ottozawa (Okuma), approximately 1 km south of FNPP, until at least 2016; thus, for invertebrates including sessile organisms in the intertidal zone near FNPP, some environmental factors seem to have inhibited reproduction, recruitment, or both.

Among the factors that might have inhibited recovery of invertebrates are strong waves; sustainability or suitability of substrate (e.g., tetrapods or similar concrete structures set along the coast for wave protection) for larvae to settle on; harmful substances such as radionuclides; heavy metals; turbidity or suspended material; insufficient amounts of prey organisms; and excess of predators. Before the FNPP accident, there were strong waves, substrate (e.g., tetrapods or similar concrete structures set along the coast for wave protection) for larvae to settle on, and various kinds of intertidal predator and prey invertebrates in the coastal waters off Fukushima Prefecture, similar to those after the FNPP accident. Although the possibility of changes in predators or prey affecting invertebrate populations cannot be ruled out, strong waves and substrate are unlikely to be those that inhibited reproduction or recruitment of invertebrates including sessile organisms in the intertidal zone near FNPP. Regarding the possible impacts by harmful substances, such as radionuclides, heavy metals and turbidity or suspended material, we need to carefully examine the effect of construction to cover the radionuclide-contaminated bottom sediments of the FNPP port; this construction was conducted between March 2012 and April 2015 [43, 44]. Approximately 33,000 m^3 of bentonite and cement were used for the construction from March 14 to July 5, 2012 [43]. Another approximately 32,500 m^3 of bentonite, sand, and cement were also used for the construction from July 17, 2014 to April 23, 2015 [44]. Cement is known to include several heavy metals, and the specific gravity of cement is 3050 kg/m^3 [45]. We tried to estimate the total amount of heavy metals, which had been deposited into the marine environment during the construction. Assuming that the half of total amount of material (i.e., bentonite, sand and cement) used for the construction was cement, then the total weight of cement used for construction to cover the radionuclide-contaminated bottom sediments of the FNPP port was approximately 100,000 metric tons. This cement included various amounts of chromium (Cr), copper (Cu), zinc (Zn) and lead (Pb) and other heavy metals, which were intentionally deposited into the marine environment adjacent to FNPP; consequently, the coastal waters of Fukushima Prefecture, especially close to FNPP, might have been contaminated (Table 6.4). Thus, these heavy metals, as well as turbidity or suspended material from the construction activity, may have had ecological effects (i.e., adverse effects on larval, juvenile, and adult invertebrates in the intertidal zone) that should be elucidated in the near future. Similar impacts on the number of species and population densities by the restoration work of coastal structures were also likely at Tomioka Fishing Port (Tomioka) (since 2015), Urajiri (since 2013) and Ishinomaki (since 2012) [46–48]. The lack of observed egg capsules spawned by rock shells at Urajiri

Table 6.4 General composition of heavy metals in cement and their estimated total amounts used in the construction to cover the radionuclide-contaminated bottom sediments of the FNPP port from March 2012 to April 2015

Metals	Composition (mg/kg)	Estimated total amount (t)
Total Cr	98	9.8
Hexavalent Cr	10.8	1.1
Cu	140	14.0
Zn	511	51.0
As	18.9	1.9
Se	<1	<0.1
Cd	2.0	0.2
Total Hg	0.023	0.002
Pb	111	11.1

General composition of metals as reported by [45]. We estimated the total amount of cement used for the construction to be approximately 100,000 metric tons, assuming that cement comprised half of the total amount of material (i.e., bentonite, sand and cement) used

in 2015 (Table 6.3) might also have been associated with the restoration of coastal structures.

There have been many scientific researches and review papers on the toxicities of heavy metals to marine and estuarine organisms [49–54]. Although it is known that the toxicities of heavy metals to marine and estuarine organisms can vary according to their chemical speciation (e.g., free ions are more toxic than other chemical species) and pH [50, 53, 54], the embryos and larvae of marine or estuarine invertebrates are generally sensitive to heavy metals: for example, LC_{50} values (the concentrations at which 50% of the tested organisms die) for the embryos of American oyster (*Crassostrea virginica*) were 5.6 ppb (µg/L), 5.8 ppb (µg/L), and 103 ppb (µg/L) for mercury (Hg), silver (Ag) and Cu, respectively. LC_{50} values for nickel (Ni), Pb and cadmium (Cd) were 1.18 ppm (mg/L), 2.45 ppm (mg/L), and 3.80 ppm (mg/L), respectively [52]. Moreover, in the presence of multiple metals, the toxicities are reported to be additive; the presence of multiple metals may be more realistic than the presence of single metals in marine/estuarine environments [54].

The effect of turbidity and suspended material in aquatic environments has been well studied (e.g., Stern and Stickle [55]). Most studies on adult estuarine and marine bivalves (clams, oysters, and mussels) have indicated that the mortality rate among populations adjacent to dredging and disposal areas is low, except for individuals directly buried by the disposal operation; however, the percentage occurrence of normally developing eggs and larvae may decrease as the concentration of suspended solids increases in the range of concentrations normally resulting from dredging and disposal [55]. Thus, further experimental studies are necessary to elucidate the effect of construction to cover the radionuclide-contaminated bottom sediments of the FNPP port, as well as other restoration work of coastal structures.

The exposure situation for intertidal organisms in Fukushima Prefecture could be complex with many aspects, including various potential direct impacts such as physical harm from the tsunami, and toxicity from chemicals and radionuclides in the massive release immediately after the accident, potentially leading to acute effects [34]. Thereafter, there could have been continued releases to the sea of radionuclides and other harmful substances (i.e., heavy metals and turbidity or suspended material from the construction to cover the radionuclide-contaminated bottom sediments of the FNPP port); these releases could have had ecological effects, for example, on interspecific relationships (such as predator-prey relationships or competition for prey organisms and habitat). The effect could also have involved intraspecific relationships (competition for prey organisms, habitat and mating partners) [56–58].

At much higher dose rates, possibly immediately after the accident, differences among taxa in sensitivity to radiation could create competitive advantages for resistant organisms within a taxon and between populations of interacting taxa [23, 59]. Thus, in addition to differences in the direct radiosensitivity of individual organisms, life history traits including responses to changes in resources and generation times affect the consequences of radiation exposure. Exposures to radiation may have been high during or shortly after the accident [56]. In addition, note that particularly at lower doses, ecological factors and variability can be more important than direct radiation effects; therefore, a different conceptual methodology may be necessary to assess ecosystem-level effects, possibly including site-specific assessment of the effect of potential disturbances on ecosystems [56].

Further studies are needed to clarify the main causal factors for declining intertidal biota and subsequent slow recovery near FNPP, possibly through determining the acute, subacute and chronic toxicity of various radionuclides, chemicals and other factors in laboratory experiments. Continued field observations of spatiotemporal changes in the populations of sessile organisms near FNPP, including rock shell populations, are also necessary to evaluate their recovery in the future; these studies should consider the characteristics of habitats that may influence the distribution of sessile organisms. The focus should be on increasing population densities and reproductive success in terms of active behaviors such as mating and egg-laying and the subsequent successful recruitment of larvae and juveniles. Both field and laboratory studies will also be necessary to observe and evaluate possible multigenerational effects such as changes in reproductive success resulting from exposure to low-dose radiation and other environmental stressors.

6.5   Conclusions

We conclude that the population densities and species richness of intertidal invertebrates along the coast off Fukushima Prefecture have decreased since March 2011, especially south of FNPP. The recovery from this decline in population densities of intertidal invertebrates, including *T. clavigera*, as well as in the number of invertebrate species, has been limited and slow at sites located near and south of FNPP; further studies are needed to clarify the main causal factors.

Acknowledgment Many thanks to the Fukushima Prefectural Government for permission to enter the restricted area around FNPP. Funding was provided by a Grants-in-Aid for Scientific Research (B) to T.H. (#15H04537) from the Japan Society for the Promotion of Science.

References

1. Steinhauser G, Brandl A, Johnson TE (2014) Comparison of the Chernobyl and Fukushima nuclear accidents: a review of the environmental impacts. Sci Total Environ 470–471:800–817
2. Tokyo Electric Power Company (2012a) Estimation of the amount of radionuclides emitted from Fukushima Daiichi Nuclear Power Plant accidents to the atmosphere. [in Japanese] Available at: http://www.tepco.co.jp/cc/press/betu12_j/images/120524j0105.pdf. Accessed 21 Mar 2018
3. Tokyo Electric Power Company (2012b) Estimation of the amount of radionuclides emitted from Fukushima Daiichi Nuclear Power Plant accidents to the sea (around the Fukushima Daiichi Nuclear Power Plant exclusive harbour). [in Japanese]. Available at: http://www.tepco. co.jp/cc/press/betu12_j/images/120524j0102.pdf. Accessed 21 Mar 2018
4. Aoyama M, Kajino M, Tanaka TY et al (2016) ^{134}Cs and ^{137}Cs in the North Pacific Ocean derived from the March 2011 TEPCO Fukushima Dai-ichi Nuclear Power Plant accident, Japan. Part two: estimation of ^{134}Cs and ^{137}Cs inventories in the North Pacific Ocean. J Oceanogr 72:67–76
5. Estournel C, Bosc E, Bocquet M et al (2012) Assessment of the amount of cesium-137 released into the Pacific Ocean after the Fukushima accident and analysis of its dispersion in Japanese coastal waters. J Geophys Res 117:C11014
6. Kawamura H, Kobayashi T, Furuno A et al (2011) Preliminary numerical experiments on oceanic dispersion of ^{131}I and ^{137}Cs discharged into the ocean because of the Fukushima Daiichi nuclear power plant disaster. J Nucl Sci Technol 48:1349–1356
7. Kobayashi T, Nagai H, Chino M et al (2013) Source term estimation of atmospheric release due to the Fukushima Dai-Ichi nuclear power plant accident by atmospheric and oceanic dispersion simulations. J Nucl Sci Technol 50:255–264
8. Smith G (2014) Sources, effects and risks of ionizing radiation. UNSCEAR 2013 report. Volume I: report to the general assembly, annex A: levels and effects of radiation exposure due to the nuclear accident after the 2011 Great East-Japan earthquake and tsunami. J Radiol Prot Off J Soc Radiol Prot 34:725
9. Charette MA, Breier CF, Henderson PB et al (2013) Radium-based estimates of cesium isotope transport and total direct ocean discharges from the Fukushima nuclear power plant accident. Biogeosciences 10:2159–2167
10. Miyazawa Y, Masumoto Y, Varlamov SM et al (2013) Inverse estimation of source parameters of oceanic radioactivity dispersion models associated with the Fukushima accident. Biogeosciences 10:2349–2363
11. Rypina II, Jayne SR, Yoshida S et al (2013) Short-term dispersal of Fukushima-derived radionuclides off Japan: modeling efforts and model-data intercomparison. Biogeosciences 10:4973–4990
12. Tsumune D, Tsubono T, Aoyama M et al (2012) Distribution of oceanic ^{137}Cs from the Fukushima Daiichi nuclear power plant simulated numerically by a regional ocean model. J Environ Radioact 111:100–108
13. Tsumune D, Tsubono T, Aoyama M et al (2013) One-year, regional-scale simulation of 137Cs radioactivity in the ocean following the Fukushima Daiichi nuclear power plant accident. Biogeosciences 10:5601–5617
14. Buesseler K, Dai M, Aoyama M et al (2017) Fukushima Daiichi-derived radionuclides in the ocean: transport, fate, and impacts. Annu Rev Mar Sci 9:173–203

15. Fisheries Agency of Japan (2018) Results of the monitoring on radioactivity level in fisheries products. Available at http://www.jfa.maff.go.jp/e/inspection/index.html. Accessed 21 Mar 2018
16. Wada T, Nemoto Y, Shimamura S et al (2013) Effects of the nuclear disaster on marine products in Fukushima. J Environ Radioact 124:246–254
17. Wada T, Fujita T, Nemoto Y et al (2016) Effects of the nuclear disaster on marine products in Fukushima: an update after five years. J Environ Radioact 164:312–324
18. International Commission on Radiological Protection (2008) Environmental protection – the concept and use of reference animals and plants. ICRP Publication 108. Ann ICRP 38:4–6. (ICRP, 2008)
19. Vandenhove H (2011) Effects of ionizing radiation on non-human biota. Available at: http://www.bvsabr.be/15april2011/Vandenhove.pdf. Accessed 30 June 2015
20. United Nations Scientific Committee on the Effects of Atomic Radiation (1996) Sources and effects of ionizing radiation, UNSCEAR Report to the General Assembly, with Scientific Annex: effects of radiation on the environment. United Nations sales publication E.96.IX.3 (United Nations, 1996)
21. United Nations Scientific Committee on the Effects of Atomic Radiation (2008) Sources and effects of ionizing radiation. UNSCEAR Report to the General Assembly, with Scientific Annexes, Annex E: effects of ionizing radiation on non-human biota. (United Nations, 2008)
22. United Nations Scientific Committee on the Effects of Atomic Radiation (2011) Sources and effects of ionizing radiation. Volume II: effects, Scientific Annexes C, D and E. United Nations Scientific Committee on the Effects of Atomic Radiation sales publication E.11.IX.3 (United Nations, 2011)
23. Copplestone D, Hingston JL, Real A (2008) The development and purpose of the FREDERICA radiation effects database. J Environ Radioact 99:1456–1463
24. FREDERICA radiation effects database. (2006). Available at: http://www.frederica-online.org/mainpage.asp. Accessed 21 Mar 2018
25. Garnier-Laplace J, Copplestone D, Gilbin R et al (2008) Issues and practices in the use of effects data from FREDERICA in the ERICA integrated approach. J Environ Radioact 99:1474–1483
26. Andersson P, Garnier-Laplace J, Beresford NA et al (2009) Protection of the environment from ionising radiation in a regulatory context (PROTECT): proposed numerical benchmark values. J Environ Radioact 100:1100–1108
27. Beresford NA, Brown J, Copplestone D et al. (2007) D-ERICA: an integrated approach to the assessment and management of environmental risks from ionising radiation. *ERICA project* (FI6R-CT-2004-508847, 2007). Available at: https://wiki.ceh.ac.uk/download/attachments/115017395/D-Erica.pdf. Accessed 21 Mar 2018
28. Miura O, Kanaya G (2017) Impact of the 2011 Tohoku earthquake tsunami on marine and coastal organisms. Biol Int S36:81–92
29. Urabe J, Suzuki T, Nishita T et al (2013) Immediate ecological impacts of the 2011 Tohoku earthquake tsunami on intertidal flat communities. PLoS One. https://doi.org/10.1371/journal.pone.0062779
30. Hidaka M, Wakui K, Kamiyama K et al (2012) Change in sediment characters and bathymetry in Matsukawaura Inlet due to the tsunami on March 11, 2011. J Jpn Soc Civ Eng, B3 (Hydraul, Coast Environ Eng) 68:I_186–I_191. [in Japanese]
31. Fukushima Prefectural Government (2015a) Surveys on the environment and aquatic organisms in Matsukawaura, Fukushima. http://www.pref.fukushima.lg.jp/uploaded/attachment/141950.pdf [in Japanese]
32. Seike K, Shirai K, Kogure Y (2013) Disturbance of shallow marine soft-bottom environments and megabenthos assemblages by a huge tsunami induced by the 2011 M9.0 Tohoku-Oki earthquake. PLoS One. https://doi.org/10.1371/journal.pone.0065417
33. Takami H, Won NI, Kawamura T (2012) 5. Ontogenetic habitat shift in ezo abalone *Haliotis discus hannai*. Nippon Suisan Gakkaishi 78:1213–1216. in Japanese

34. Horiguchi T, Yoshii H, Mizuno S et al (2016) Decline in intertidal biota after the 2011 Great East Japan earthquake and tsunami and the Fukushima nuclear disaster: field observations. Sci Rep 6:20416. https://doi.org/10.1038/srep20416
35. Clarke KR (1993) Non-parametric multivariate analyses of changes in community structure. Australian J Ecol 18:117–143
36. Clarke KR, Warwich RM (2001) Change in marine communities: an approach to statistical analysis and interpretation, 2nd edn. PRIMER-E, Plymouth
37. Tokyo Electric Power Company (2001) Marine organisms. In: Environmental impact assessment, regarding the construction of Units 7 and 8 of the Fukushima Daiichi Nuclear Power Plant, 5.11-1-5.11-14 (TEPCO, 2001). [in Japanese]
38. Nakano D, Nagoshi M (1980) Growth and age of *Thais clavigera* (Küster), Prosobranch, in tidal zone around Shima Peninsula, Japan. 25th anniversary memorial Journal for Toba Aquarium, 87–92. [in Japanese]
39. Kubota M, Nakata K, Nakamura Y (1981) Continental shelf waves off the Fukushima coast part I: observations. J Oceanogr Soc Jpn 37:267–278
40. Kaeriyama H (2017) Oceanic dispersion of Fukushima-derived radioactive cesium: a review. Fish Oceanogr 26:99–113
41. Yasuda I (2003) Hydrographic structure and variability in the Kuroshio-Oyashio transition area. J Oceanogr 59:389–402
42. Horiguchi T, Kodama K, Aramaki T et al. (2018) Radiocesium in seawater, sediments, and marine megabenthic species in coastal waters off Fukushima in 2012–2016, after the 2011 nuclear disaster. Mar Environ Res, in press
43. Tokyo Electric Power Company (2012c) Construction to cover the bottom sediments of FNPP specific port (as of July 18, 2012). [in Japanese]. Available at: http://www.tepco.co.jp/nu/fukushima-np/images/handouts_120718_02-j.pdf. Accessed 21 Mar 2018
44. Tokyo Electric Power Company (2015) Construction to cover the bottom sediments of FNPP specific port (as of April 23, 2015). [in Japanese]. Available at: http://www.tepco.co.jp/nu/fukushima-np/handouts/2015/images/handouts_150423_02-j.pdf. Accessed 21 Mar 2018
45. Ugajin T (2001) The effects of the trace elements in cement on the environment. Concr J 39:14–19. [in Japanese]
46. Fukushima Prefectural Government (2015b) Reconstruction of Tomioka Fishing port (as of September 24, 2015). [in Japanese]. Available at: https://www.pref.fukushima.lg.jp/sec/41390a/tomioka-kaigan.html. Accessed 21 Mar 2018
47. Reconstruction Agency (2015a) Activities plan for reconstruction of Minami-Soma City, Fukushima Prefecture. (as of July 31, 2015). [in Japanese]. Available at: http://www.reconstruction.go.jp/topics/main-cat1/sub-cat1-3/20150731/20150731_Fukushima03Minamisoma.pdf. Accessed 21 Mar 2018
48. Reconstruction Agency (2015b) Activities plan for reconstruction of Ishinomaki City, Miyagi Prefecture. (as of July 31, 2015). [in Japanese]. Available at: http://www.reconstruction.go.jp/topics/main-cat1/sub-cat1-3/20150731/20150731_Miyagi03Ishinomaki.pdf. Accessed 21 Mar 2018
49. Ansari TM, Marr IL, Tariq N (2004) Heavy metals in marine pollution perspective–a mini review. J Appl Sci 4:1–20
50. Bryan GW (1971) The effects of heavy metals (other than mercury) on marine and estuarine organisms. Proc R Soc Lond B Biol Sci 177:389–410. https://doi.org/10.1098/rspb.1971.0037
51. Bryan GW, Langston WJ (1992) Bioavailability, accumulation and effects of heavy metals in sediments with special reference to United Kingdom estuaries: a review. Environ Pollut 76:89–131
52. Calabrese A, Collier S, Nelson DA et al (1973) The toxicity of heavy metals to embryos of the American oyster *Crassostrea virginica*. Mar Biol 18:162–166
53. Rainbow PS (1985) The biology of heavy metals in the sea. Int J Environ Stud 25:195–211. https://doi.org/10.1080/00207238508710225

54. Verslycke T, Vangheluwe M, Heijerick D et al (2003) The toxicity of metal mixtures to the estuarine mysid *Neomysis integer* (Crustacea: Mysidacea) under changing salinity. Aquat Toxicol 64:307–315
55. Stern EM, Stickle WB (1978) Effects of turbidity and suspended material in aquatic environments: literature review, Technical Report D-78-21 (Dredged Material Research Program) 117 p. U. S. Army Engineer Waterways Experiment Station, Vicksburg
56. Bradshaw C, Kapustka L, Barnthouse L et al (2014) Using an ecosystem approach to complement protection schemes based on organism-level endpoints. J Environ Radioact 136:98–104
57. Bréchignac F, Bradshaw C, Carroll S et al (2011) Recommendations from the international union of radioecology to improve guidance on radiation protection. Integr Environ Assess Manag 7:411–413
58. International Union of Radioecology (2013) Towards an ecosystem approach for environment protection with emphasis on radiological hazards. IUR Report no.7 – 2nd edn. (978-0-9554994-4-9, 2012). Available at: www.iur-uir.org/en/publications/others-publications/id-15-iur-report-7-towards-an-ecosystem-approach-for-environment-protection-with-emphasis-on-radiological-hazards. Accessed 21 Mar 2018
59. Whicker WF, Schultz V (1982) Radioecology: nuclear energy and the environment, vol 1. CRC Press, Boca Raton, p 212

7

mtDNA Mutations in Mano River Salmon

Muhammad Fitri Bin Yusof, Gyo Kawada, Masahiro Enomoto,
Atsushi Tomiya, Masato Watanabe, Daigo Morishita, Shigehiko Izumi,
and Masamichi Nakajima

Abstract On March 11, 2011, a great earthquake occurred off the east coast of Honshu Island, Japan. The consequent breakdown of the Fukushima Daiichi Nuclear Power Plant (FDNPP) caused a massive release of radionuclides into terrestrial and marine environments and into the atmosphere. The Abukuma Mountains region is one of the areas highly polluted by this accident. Freshwater fishes continued to live in this area after the FNPP accident became so polluted.

Mitochondrial DNA (mtDNA) usually transfers from the mother to the next generation clonally. Therefore, it is one of the best genetic predictors of the effect of radiation on DNA. The influence of radiation can be presumed by comparing mtDNA between larval fish and their female parent. mtDNA of masu salmon was collected from an area highly polluted, namely, the upstream portion of Mano River, and compared it with that from nonpolluted cultured masu salmon.

While no mutations were observed in the cultured masu salmon, those collected from Mano River exhibited three types of subdivisions in the Cytb region and two types of subdivisions in the D-loop region of mtDNA. These results suggest that exposure to radioactive cesium causes a base exchange in DNA. But the mutations observed were not serious enough to affect the masu salmon phenotype.

Keywords Mitochondrial DNA · Mutation · Cytb · D-loop · *Oncorhynchus masou* · Fukushima

M. F. B. Yusof · M. Nakajima (✉)
Graduate School of Agricultural Science, Tohoku University, Sendai, Japan
e-mail: Masamichi.nakajima.b6@tohoku.ac.jp

G. Kawada · M. Enomoto · A. Tomiya · M. Watanabe · D. Morishita · S. Izumi
Fukushima Inland Water Fisheries Experimantal Station, Inawashiro, Fukushima, Japan

7.1   The Fukushima Daiichi Nuclear Power Plant Accident

On March 11, 2011, a great earthquake measuring 9.0 on the Richter scale occurred off the east coast of Honshu Island, Japan. It caused a huge tsunami, which struck eastern Japan and broke down the cooling system of the reactors at Fukushima Daiichi Nuclear Power Plant (FDNPP). The breakdown caused a massive release of radionuclides into terrestrial and marine environments as well as into the atmosphere. This raised a huge concern globally about how far the radionuclides would travel around the world, and their biological influences on living organisms. It also increased uneasiness among people about the long-term effect of radionuclides on ecosystems and on the next generation of humans. Precise information on what occurred and on what is ongoing has yet to be investigated. The collapse of FDNPP caused a massive release of radioactive materials into the environment [1–3]. Various potential threats to ecosystems and to living organisms were reported [4, 5]. The released radioactivity can affect an organism's genetic structure and ecosystem. More importantly, radionuclides will accumulate in the bodies of organisms and thus will be transmitted through the food chain.

The main radioactive materials that affect organisms are cesium-134 (^{134}Cs) and cesium-137 (^{137}Cs), because these materials have comparatively long half-lives of 2 and 30 years, respectively. During the decay process, these materials emit γ-rays and β-particles [6]. In addition, exposure to high-dose radioactive Cs is clastogenic to chromosomes, increases mutation rates and can be transmitted to the next generation [7, 8].

7.2   The Genetic Effects of Radiation on Organisms

Various effects of radiation have been examined since radiation was discovered. Since Muller [9] suggested that X-rays cause mutations in *Drosophila*, it has been well known that X-rays and other forms of radiation lead to genetic mutations. Casarett [10] reported the influence of radiation on mammals such as mice. In the case of fishes, many studies have examined the genetic effects of X-rays and γ-rays from the 1960s to the 1980s [11–13] in guppy (*Poecilia reticulata*) and medaka (*Oryzias latipes*). These studies used relatively high dose of external radiation. On the other hand, in the case of the FDNPP accident, it is necessary to consider the effect of internal low-dose and long-term exposure to radioactive substances. Little is known about the effect of long-term exposure to low levels of radioactive substances such as ^{137}Cs. The effect of ^{137}Cs contamination on organisms should be elucidated, especially with regard to changes in chromosome structure, gametes, molecular structure and DNA sequence.

7.3 The Situation of Freshwater Fishes in Fukushima

A large amount of radioactive Cs was released into the atmosphere as a result of the FDNPP accident in March 2011. Chino et al. [14] estimated the total amount of released radioactive materials until April 2011, as follows: iodine-131 (^{131}I) was 1.5 $\times$ 10^{17} Bq and ^{137}Cs was 1.2×10^{16} Bq. On the other hand, Stohl et al. [15] estimated the released radioactive materials as follows: Xenon-133 (^{133}Xe) was 1.24×10^{19} Bq and ^{137}Cs was 2.94×10^{18} Bq. Radioactive Cs travels through both air and water systems, and in this case, it contaminated most aquatic organisms across a vast area [1, 16, 17]. The Abukuma Mountains area is one of the areas highly polluted by the accident. Radioactive pollution in the freshwater area in the Abukuma Mountains continues to the present [18–20] (as of August, 2019, http://fikushima-radioactivity.jp/pc/). Freshwater fishes have continued to live in this polluted environment after the accident.

Masu salmon (*Oncorhynchus masou*), belongings to the salmonid family of genus *Oncorhynchus*, is widely distributed in Russia, Korea, and Japan, including the Abukuma Mountains. In Japan, they are distributed mainly in the coastal area of the Sea of Okhotsk, the Northern Pacific Ocean and the Northern Sea of Japan [21, 22]. However, masu salmons are found in all areas of Mano River where sampling was performed in this study; their distribution between upstream and downstream is separated by Hayama dam. The upstream salmons are separated from the downstream population and spend their life cycle in the dam lake and the upper stream. The individuals collected from Hayama Lake and its upper stream had spent 2–3 years in this ecosystem polluted by radionuclides from the FDNPP accident.

7.4 Use of Gynogenesis to Estimate Mutation Rate

Mitochondrial DNA (mtDNA) is one of the best genetic predictors of the effect of radiation on DNA because mtDNA usually transfers to the next generation clonally. mtDNA also has a higher mutation rate than genomic DNA [23]. It is expected that mtDNA is more sensitive to radiation than genomic DNA. Because a cell contains thousands of copies of mitochondria, there is a possibility of heteroplasmy. But, since it was reported that heteroplasmic mtDNA quickly return to homoplasmy, it can be assumed that the fundamental haplotype of each individual is homoplasmic and monomorphic [24]. Gynogenesis involves the production of diploid eggs in which both chromosome sets are obtained from the maternal half. An example of paternal inheritance was also reported, but it is a very rare occurrence [25]. The influence of mtDNA from the male parent can be excluded more certainly by using artificial gynogenesis technology. It is expected that the offspring obtained from a single female will possess basically the same DNA sequence. Fitri et al. (unpublished) compared the sequences of two regions of mtDNA of offspring obtained from two females collected from Mano River in Fukushima Prefecture against those collected from two cultured females from the Fukushima Prefectural Inland Water Fisheries Experimental Station (FPIWFES).

The application of artificial gynogenesis technology to fish was established in the 1980s [26]. This technology was used to examine the production of tetraploid, polyploid and clonal fishes, and these attempts succeeded in many fishes and shellfishes. The clonalization of mtDNA in masu salmon using artificial gynogenetic technology is not difficult. The influence of radiation can be presumed by examining what happens in the mtDNA introduced from a dam as a clone larval fish.

7.5   Genetic Analysis in This Study

7.5.1   Materials and Methods

Masu salmons that had swum upstream to spawn were collected from Mano River upstream of Hayama Lake (Fig. 7.1). The fishes examined in this study were some of those individuals. We succeeded in obtaining gynogenetic offspring from two females from Mano River and two cultured females from the FPIWFES.

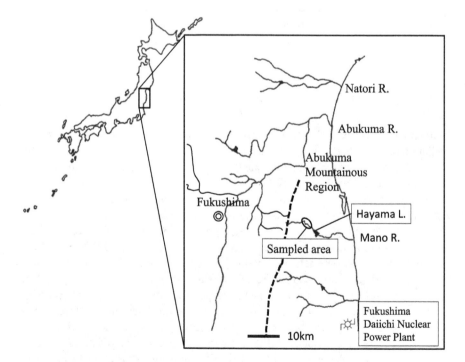

Fig. 7.1 Locations of sample collection, Mano River, Fukushima

Table 7.1 Used microsatellite DNA markers in this study and their forward and reverse primers

Locus	Repeat sequence of cloned allele	Primer sequence (5'–3') (F, forward; R, reverse)	GenBank no.
One102	(ATCT)10	F: CATGGAGAAAAGACCAATCA	AF274518
		R: TCACTGCCCTACAACAGAAG	
One114	(TAGA)12N4	F: TCATTAATCTAGGCTTGTCAGC	AF274530
	(TAGA)12	R: TGCAGGTAAGACAAGGTATCC	
Oke308	(AGAT)19	F: AAGAGCAGAGGGAGAAATGGA	
		R: GTTGTTTGGATGGGAAGTGG	

DNA was extracted from the tail fin using a DNA extraction kit (DNeasy Blood and Tissue Kit, Qiagen). Extracted DNA was used for parentage analysis and for the detection of mutations at two regions of mtDNA: Cytb and D-loop. For parentage analysis, three microsatellite primer sets were used: *One102*, *One114* [27] and *Oke308* (Kudo et al. unpublished). The primer set of each microsatellite locus is presented in Table 7.1. Primer sets corresponding to Cytb and D-loop were synthesized with reference to the mtDNA sequence of NCBI database. The forward and reverse primers of Cytb were 5'-GCCATGCACTACACCTCAGA-3' and 5'-GTTCTACGGGTATGCCTCCG-3'. The primers of D-loop were 5'-ACATCAGCACAACTCCAAGGT-3' and 5'-CGGTGCCAGGTGCTGTTA-3'. Forward and reverse sequencing analyses were performed using the 3500XL Genetic Analyzer (Life Technologies).

7.5.2 Results

7.5.2.1 Parentage Analysis

The identified genotypes of the female parent and the segregation of the offspring are presented in Table 7.2. Each maternal individual, M5 and M6, was identified by three loci of microsatellite markers, because these two maternal individuals had no common allele in the three loci examined. In *One102*, the genotype of M5 was *290/242*; that of M6 was *262/225*. The expected genotypes of the offspring in M5 were *290/290* and *242/242*, and those of the offspring in M6 were *262/262* and *225/225*. In the offspring, only the alleles that existed in the maternal parent were observed. This phenomenon was observed also in the other two microsatellite markers. These results exclude the possibility that the individual's mixture came from another family.

Table 7.2 Segregated genotypes observed in each locus of each family

Locus	Family	Maternal *genotype*	Segregated *genotype*	No. of individuals
One102	M5	*290/242*	*290/290*	9
			242/242	5
	M6	*262/225*	*262/262*	11
			225/225	5
One114	M5	*313/217*	*313/313*	7
			217/217	6
	M6	*281/253*	*281/281*	13
			253/253	4
Oke308	M5	*246/246*	*246/246*	5
	M6	*290/234*	*290/290*	8
			234/234	11

Table 7.3 Samples used in this study

Collected area	Female parent	Number of examined individuals	Age	$^{134}Cs^a$	$^{137}Cs^a$
Mano River	M5	20	2	156.0	476.0
	M6	22	1	62.6	149.0
FPIWFES	FP9	50	2	0	0
	FP23	50	2	0	0

^{a}Bq/kg

FPIWFES Fukushima Prefectural Inland Water Fisheries Experimental Station

7.5.2.2 Observed Mutations and Estimated Mutation Rate in Mitochondrial DNA of Masu Salmon

A total of 42 offsprings from Mano River, 20 from M5 and 22 from M6, were examined (Table 7.3). The concentrations of radioactive Cs in muscle tissues of M5 and M6 females were 632.0 and 211.6 Bq/kg, respectively. On the other hand, radioactive Cs in muscle tissue was undetectable in all of 100 offsprings from the FPIWFES (50 from FP9 and 50 from FP23). Although an estimated 800 bases were acquired from direct sequencing, only 765 bases were analyzed (bases 207–972).

In the Cytb region, four types of substitutions were observed. In the offspring of M5, substitutions were observed at the 324th, 697th, and 795th bases (Table 7.4). Among the four types of substitutions, amino acid substitution was observed in two types, Ala to Thr and Thr to Ala, caused by G to A and A to G base substitutions, respectively. In the M6 offspring, base substitutions were observed in the same positions as in the M5 offspring. In the D-loop region, substitutions were observed only in two M6 offspring. In these two offspring, two types of substitutions were observed at the 299th and 623rd bases, which were G to A and T to C, respectively (Table 7.5). It is possible to consider these base substitutions are caused by exposure to radioactive Cs. The individuals observed substitution have the alleles which from maternal

Table 7.4 Mutations observed in the Cytb region

Female parent	Base position	Type of substitution	Amino acid change	Observed number of individuals	Mutation rate
M5	324	T > A	Leu > Leu	1	8.04×10^{-6}
M6	324	T > G	Leu > Leu	1	1.29×10^{-5}
M5	697	G > A	Ala > Thr	1	8.04×10^{-6}
M6	697	A > G	Thr > Ala	1	4.9×10^{-6}
M5	795	G > A	Pro > Pro	2	3.1×10^{-5}
M6	795	A > G	Pro > Pro	1	4.9×10^{-6}

Table 7.5 Mutations observed in the D-loop region

Female parent	Base position	Type of substitution	Observed number of individual	Mutation rate
M6	299	G > A	2	1.76×10^{-5}
M6	623	T > C	2	1.76×10^{-5}

individual in microsatellite analysis, can exclude the possibility of contamination from other individuals. No mutation was observed in the samples from the captive maternal parents of PF9 and PF23. The Cytb region is a coding region in mtDNA. It codes for protein subunit cytochrome C complex III. The point mutation caused amino acid substitution (i.e., Ala > Thr) or just synonymous mutation. A peculiar phenomenon is the similarity of bases where mutation occurs in the Cytb region of the M5 and M6 gynogens. Since mutation is expected to be random, the occurrence of mutations of similar bases in the Cytb region might be due to the sensitivity of certain regions of Cytb to external stress. Under natural conditions, nonsynonymous mutation or deletion occurs in mtDNA due to environmental stresses [28]. Haplotype variation is another possible cause of mutations in similar bases between the M5 and M6 groups. At the 697th base, the M5 offspring has a G to A mutation, whereas those of M6 A to G mutation. This indicates that the mutation in base 697 is more likely to be the result of a haplotype variation between the M5 and M6 maternal parents.

The mutation rate at each site was calculated by using a formula by Haag-Liautard et al. [29]:

$$\mu = \sum_i \left(di \,/\, tb \right)$$

where μ is the mutation rate, di is the rate of individual observed mutations in the examined individuals, b is the total number of bases of examined individuals, and t is the number of generations (in this case, 1).

The calculated mutation rates were from 4.9×10^{-6} to 3.1×10^{-5} in the Cytb region and 1.8×10^{-5} in the D-loop region. These values were higher than the rate calculated in *Drosophila*, 6.3×10^{-8} to 9.2×10^{-8} (Haag-Liautard et al. 2008).

Mutation rates more than two orders of magnitude higher were observed in masu salmon from Mano River near to FDNPP. These results suggest that exposure to radioactive Cs causes base exchanges of DNA. However, the observed mutations do not induce any that would cause severe results such as death. Each sequence observed in this study has been registered in the DNA database even if these base substitutions led to amino acid substitutions (GenBank accession numbers AB291986, AF125210, AF429776, LC098721, LC100136, and LC100137). Gynogenesis removes lethal genes, as a homozygous individual with a lethal gene cannot survive the early stage of development, whereas a nonlethal mutation can be carried by surviving individuals. Evidence of a mutation in the Cytb region of larvae from the wild might be a synonymous mutation that does not affect any morphological function despite the amino acid change. However, since the larval phenotype has not been investigated in detail, there is no strong evidence to expect that radioactive cesium contamination in eggs does not cause morphological changes in offspring. Of course, we did not examine the sequences in individuals that could not survive until the swim-up stage. It is necessary to examine individuals in different stages of development.

7.6 Future Studies

The effect of radioactive Cs on genetic materials have not been sufficiently studied. Estimation of the mutation rate of aquatic organisms in natural conditions will be important in determining the effect of radiation on the ecosystem. The present study is the first attempt to estimate the mutation rate in an aquatic organism in an area contaminated with radionuclides from FDNPP.

High levels of radioactive Cs are still observed in the rivers of the Abukuma Mountains of Fukushima. Freshwater fishes have continued to live in these areas after the accident occurred and the situation will continue for a long time. The Japanese Government decided to continue the use of nuclear power plants in the future. This decision suggests that an in-depth response to a nuclear accident is needed. Therefore, it is necessary to monitor living conditions such as radioactive Cs concentration in this area, as well as the health conditions of fishes that live there.

Acknowledgments We are grateful to the staff members of the Fukushima Prefectural Inland Water Fisheries Experimental Station for their help with sample processing and collection. We thank all of the fishery workers in Fukushima Prefecture who helped us to collect masu salmon. This work was supported in part by JSPS KAKENHI Grant Number 25252035.

References

1. Honda MC, Kawakami H, Watanabe S et al (2013) Fukushima-derived radiocesium in western North Pacific sediment traps. Biogeosci Discuss 10:2455–2477. https://doi.org/10.5194/bgd-10-2455-2013

2. Kumamoto T, Aoyama M, Hamajima Y et al (2015) Impact of Fukushima-derived radio-cesium in the western North Pacific Ocean about ten months after the Fukushima Dai-ichi nuclear power plant accident. J Environ Radioact 140:114–122. https://doi.org/10.1016/j.jenvrad.2014.11.010

3. Masson O, Ieringer J, Barttich E et al (2016) Variation in airborne 134Cs, 137Cs, particu-late 131I and 7Be maximum activities at high-altitude European locations after the arrival of Fukushima-labeled air masses. J Environ Radioact 162:14–22. https://doi.org/10.1016/j.jenvrad.2016.05.004

4. Shigenobu T, Fujimoto K, Ambe D et al (2014) Radiocesium contamination of greenlings (Hexagrammos otakii) off the coast of Fukushima. Nat Sci Rep 4:6851. https://doi.org/10.1038/srep06851

5. Tateda Y, Tsumune D, Tsubono T et al (2015) Radiocesium biokinetics in olive flounder inhabiting the Fukushima accident-affected Pacific coastal waters of eastern Japan. J Environ Radioact 147:130–141. https://doi.org/10.1016/j.jenvrad.2015.05.025

6. Yoshihara T, Matsumura H, Hashida SN et al (2013) Radiocesium contaminations of 20 wood species and corresponding gamma-ray dose rates around the canopies at 5 months after the Fukushima nuclear power plant accident. J Environ Radioact 115:60–68. https://doi.org/10.1016/j.jenvrad.2012.07.002

7. da Cruz AD, de Melo e Silva D, da Silva CC et al (2008) Microsatellite mutations in the off-spring of irradiated parents 19 years after the Cesium-137 accident. Mutat Res/Genet Toxicol Environ Mutagen 652(2):175–179. https://doi.org/10.1016/j.mrgentox.2008.02.002

8. Kamiguchi T, Tateno H (2002) Radiation- and chemical-induced structural chromosome aber-rations in human spermatozoa. Mutat Res/Fundam Mol Mech Mutagen 504:184–191. https://doi.org/10.1016/S0027-5107(02)00091-X

9. Muller HJ (1922) X-ray induced mutation of Drosophila virilis. Science 66:84–87

10. Casarett AP (1968) Radiation biology. Prentice-Hall, New Jersey

11. Schröder JH (1969) X-ray-induced mutations in the poeciliid fish, Lebistes reticulatus Peters. Mutat Res 7:75–90

12. Hyodo-Toguchi Y, Egami N (1969) Changes in dose-survival time relationship after X-irradiation during embryonic development in the fish, Oryzias latipes. J Radiat Res 10:121–125

13. Kikuchi S, Egami N (1983) Effects of r-irradiation on the rejection of transplanted scale mela-nophores in the teleost, Oryzias latipes. Dev Comp Immunol 7:51–58

14. Chino M, Nakayama H, Nagai H et al (2011) Preliminary estimation of release amounts of 131I and 137Cs accidentally discharged from the Fukushima Daiichi nuclear power plant into the atmosphere. J Nuclear Sci Tech 48:1129–1134

15. Stohl A, Seibert P, Wotawa G et al (2012) Xenon-133 and caesium-137 releases into the atmo-sphere from the Fukushima Dai-ichi nuclear power plant: determination of the source term, atmospheric dispersion, and deposition. Atmos Chem Phys 12:2313–2343

16. Povinec PP, Gera M, Holy K et al (2013) Dispersion of Fukushima radionuclides in the global atmosphere and the ocean. Appl Tadiat Isot 81:383–392

17. Zaleaska T, Suplinska M (2013) Fish pollution with anthropogenic 137Cs in the southern Baltic Sea. Chemosphere 90:1760–1766. https://doi.org/10.1016/jchemosphere.2012.07.012

18. Iwagami S, Tsujimura M, Onda Y et al (2017) Contribution of radioactive 137Cs discharge by suspended sediment, coarse organic matter, and dissolved fraction from a headwater catch-ment in Fukushima after the Fukushima Dai-ichi nuclear power plant accident. J Environ Radioact 166:466–474. https://doi.org/10.1016/j.jenvrad.2016.07.025

19. Yoshimura M, Yokoduka T (2014) Radioactive contamination of fishes in lake and streams impacted by the Fukushima nuclear power plant accident. Sci Total Environ 482–483:184–192. https://doi.org/10.1016/j.scitotenv.2014.02.118

20. Wada T, Tomiya A, Enomoto M et al (2016) Radiological impact of the nuclear power plant accident on freshwater fish in Fukushima: an overview of monitoring results. J Environ Radioact 151:144–155. https://doi.org/10.1016/j.jenvrad.2015.09.017

21. Kitanishi S, Edo K, Yamamoto T et al (2012) Fine scale relationships between sex, life history, and dispersal of masu salmon. Ecol Evol 2:920–929. https://doi.org/10.1002/ece3.228. et al 2012

22. Ohkuma K, Ishida Y, Rassadnikov OA et al (2000) Distribution and biological characters of pink (*Oncorhynchus gorbuscha*) and masu salmon (*O. masou*) in the Sea of Japan. Bull Natl Salmon Resour Cent 3:1–10

23. Wang L, Kuwahara Y, Li L et al (2007) Analysis of common deletion (CD) and a novel deletion of mitochondrial DNA induced by ionizing radiation. Int J Radiat Biol 83:433–442

24. Ashley MV, Laipis PJ, Hauswirth WW (1989) Rapid segregation of heteroplasmic bovine mitochondria. Nucleic Acids Res 17:7325–7331

25. Schwartz M, Vissing J (2002) Paternal inheritance of mitochondrial DNA. N Engl J Med 347:576–580

26. Thorgaard GH, Allen SK (1986) Chromosome manipulation and markers in fishery management. In: Ryman N, Utter F (eds) Population genetics and fishery management. Washington Sea Grant Program, Distributed by University of Washington Press, Seattle, pp 319–331

27. Olsen JB, Wilson SL, Kretschmer EJ, Jones KC, Seeb JE (2009) Characterization of 14 tetranucleotide microsatellite loci derived from sockeye salmon. Mol Ecol 9:2155–2234

28. Shirai K, Inomata N, Mizoiri S et al (2014) High prevalence of non-synonymous substitutions in mtDNA of cichlid fishes from Lake Victoria. Gene 552:239–245. https://doi.org/10.1016/j.gene.2014.09.039

29. Haag-Liautard C, Coffey N, Houle D et al (2008) Direct estimation of the mitochondrial DNA mutation rate in *Drosophila melanogaster*. PLoS Biol 6(8):1706–1714. https://doi.org/10.1371/journal.pbio.0060204

8

Coniferous trees after the
FNPP accident

Yoshito Watanabe

Abstract After the accident at Fukushima Daiichi Nuclear Power Plant (FNPP) in March 2011, high contamination levels in the environment suggest possible effects of radiation on nonhuman biota. In order to understand the effect on wild animals and plants, field investigations were conducted in the ex-evacuation zone where ambient dose-rate was particularly high. For the purpose of biomonitoring of the radiation effect, coniferous trees have been demonstrated to be suitable indicator organisms because of their high radiosensitivity, which was revealed decades ago by experiments using gamma irradiation facilities. Subsequently, radiosensitive damages in conifers became real after the Chernobyl nuclear power plant (CNPP) accident in 1986, where local coniferous species showed distinct biological damage in the radioactively contaminated areas. This review outlines the results obtained from the past radiation experiments and cases in surrounding forests after the CNPP accident for radiation effect studies of coniferous trees. By referring to them, the author explains the present situation and problems of the investigation on coniferous trees after the FNPP accident.

Keywords Radiation effect · Radiosensitive plants · Coniferous trees

8.1 Introduction

Following the accident at Fukushima Daiichi Nuclear Power Plant (FNPP) in March 2011, there are concerns about radionuclide contamination in the forests around FNPP. In addition to the distribution and dynamics of radionuclides in the forest environment, much attention has been paid to the biological consequences of radiation emitted from radionuclides on animals and plants that inhabit the forest. To

Y. Watanabe (✉)
Fukushima Project Headquarters, National Institute of Radiological Sciences, National Institutes for Quantum and Radiological Science and Technology, Chiba, Japan
e-mail: watanabe.yoshito@qst.go.jp

understand the impact on such wild animals and plants, examinations have been carried out in heavily contaminated areas within the ex-evaluation zone where "it is expected that the residents have difficulties in returning for a long time (Ministry of Economy, Trade and Industry)" since 2011 [1]. Some of the examinations were made on coniferous tree as a possible indicator organism of the radiation effect, because this group of plants has been evaluated to have radiosensitive characteristics.

With regard to the radiation effect in coniferous trees, research that takes environmental impacts into consideration has been conducted in the past. These were conducted primarily as irradiation experiments using outdoor γ-ray irradiation facilities (gamma fields). Many gamma fields existed in Europe and the United States (US) from the 1950s to the 1970s, with the first gamma field established in the US Brookhaven National Laboratory in 1948 [2]. In Japan, research was focused on the current gamma field at the Radiation Breeding Division at the National Agriculture and Food Research Organization [3]. In the gamma field of US and Canada, the changes in trees were monitored for ten or more years after placing a cesium (Cs) radiation source within a forest [4, 5]. However, a conventional study on radiation biology that primarily focused on observing the growth and morphological changes in coniferous trees has been almost complete, and since then, the current number of radiation biology researches on coniferous trees in Europe and US has been declining rapidly.

Meanwhile, unlike in Western Europe, US and Japan, radiation impact research on coniferous trees continued to take place beyond the 1980s in USSR. In the accident at the Mayak Production Association in 1957 and the Chernobyl Nuclear Power Plant (CNPP) accident in 1986, there were actual radiation injuries in coniferous trees within the natural environment and many studies investigating the impact of the accidents. In recent years, due to the growing interest in environmental protection, the influence of radiation on the living environment has begun to gain global attention. This has led to efforts on the issue by international authorities that included publication of reports reviewing the investigations into the radiation effect on forest plants in USSR. Beginning with the 1994 report by the United Nations Scientific Committee on the Effects of Atomic Radiation (UNSCEAR) [6], the accident report of the International Atomic Energy Agency (IAEA) on the CNPP accident provides a detailed review on the impact of radiation on coniferous trees, to demonstrate how organisms in the environment have been affected [7].

This review outlines the results obtained from irradiation experiments in the past and the cases observed in the surrounding forests after the CNPP accident, as radiation impact research of coniferous trees. By referring to these, the author explains the present situation and issues associated with the investigation on coniferous trees following the FNPP accident.

8.2   Investigations Using Irradiation Facilities on Coniferous Trees

8.2.1   Radiation Hypersensitivity Observed at the Lethal Dose/ Dose-Rate on Coniferous Trees

Higher plant species show great differences in radiation sensitivity, which has been demonstrated by researches in US using gamma fields. Differences in radiosensitivity of various plant species were compared for growth inhibition and mortality after acute radiation exposure [8, 9]. In woody plants, 28 species (12 broad-leaved and 16 coniferous trees) were studied on the lethal dose 2 years after acute irradiation [9]. Although the lethal dose in the broad-leaved trees spanned a wide range between 2 and 20 kR (20 and 200 Gy), the coniferous trees showed distinctively lower lethal doses between 0.8 and 1.6 kR (8 and 16 Gy).

As in acute irradiation experiments, hypersensitivity of coniferous trees to radiation was observed also in chronic irradiation experiments [10, 11]. The mean lethal dose-rate after irradiating six species of coniferous trees for 3 years was 22.9 R/day (229 mGy/day), which was approximately 10% of the mean lethal dose-rate in seven species of broad-leaved trees. Chronic irradiation experiments in the Japanese gamma field have also been conducted on Japanese coniferous trees such as Japanese cedar (*Cryptomeria japonica*), Japanese cypress (*Chamaecyparis obtusa*), Japanese red pine (*Pinus densiflora*) and Japanese black pine (*Pinus thunbergii*), with the lethal dose-rate between 10 and 20 R/day (100 and 200 mGy/day) having been observed after 2 years of irradiation [12].

Furthermore, coniferous trees exposed to radiation over an extended period of time tend to accumulate radiation injuries, and the lethal dose-rate declines with the irradiation period [4]. In pitch pine (*Pinus rigida*), as γ-irradiation was extended from 8 to 10 years, the lethal dose-rate decreased from >10.6 R/day (over 106 mGy/ day) to 5.3 R/day (53 mGy/day), while the 50% lethal dose-rate decreased from 3.1 R/day (31 mGy/day) to 3.0 R/day (30 mGy/day) [4]. In balsam fir (*Abies balsamea*), which is thought to be particularly sensitive to radiation, the 50% lethal dose-rate with 11 years of γ-irradiation decreased down to 28.5 mGy/day and tended to decrease even further as the irradiation period was extended [5].

Radiation hypersensitivity seen in coniferous trees corresponds to cellular traits unique to this plant group. Coniferous trees have a wide variety of species (about 500 species) from the temperate zone to the subarctic zone around the world, but what is common to them all is a very large genome size (the amount of DNA per genome) [13], which results in a very large volume of cell nucleus. Therefore, even when exposed to radiation of the same absorbed dose, the energy of radiation absorbed per cell nucleus or one chromosome in coniferous trees is larger than that of a broad-leaved tree with smaller genome size and cell nucleus volume. To that end, Sparrow et al. explained that the susceptibility of coniferous trees to radiation influences at the cellular level is responsible for radiation sensitivity of coniferous trees at the organismal level [8, 9, 14].

8.2.2 Various Radiation Injuries in Coniferous Trees

In plants, various radiation injuries also occur when they are subjected to chronic irradiation below the lethal dose-rate. The minimum dose-rate at which such radiation injury occurs correlates positively with the lethal dose-rate of the plant species. Accordingly, coniferous trees with the low lethal dose-rate can be injured even at a relatively low exposure dose-rate [15]. In pitch pine, the radial growth of the tree trunk decreased after several years of irradiation at 4 R/day (40 mGy/day) [4], and the length of needles was halved after 10 years of irradiation at a mean dose-rate of 3 R/day (30 mGy/day), while minor shrinkage of needles was observed even at 1.5 R/day (15 mGy/day) of irradiation [4]. A decrease in tree growth was obviously detectable in pitch pine within several years of irradiation at 2 R/day (20 mGy/ day) [14].

In particular, the injury that is noticeably observed even at low dose-rates is the growth of the tip of the tree shoot (apical growth), associated with cell division and elongation in apical meristematic tissues, and formation of buds and reproductive organs. The annual growth in height of Japanese larch (*Larix kaempferi*) begun to be suppressed at 2 R/day (20 mGy/day), and its 50% reduced dose-rate was estimated to be 2.5–4.5 R/day (25–45 mGy/day) [16]. In pitch pine, the higher the dose-rate, the lower the growth of leader shoot, the greater the growth of lateral sprouts, and the lower was the number of lateral buds formed newly [17]. In balsam fir, the amount of shoot elongation and the number of buds formed were significantly decreased by γ-irradiation of 19 mGy/day for 3 years [5]. Furthermore, in Taxus media (*Taxus media*), the number of buds formed decreased significantly after irradiation at 3.75 R/day (37.5 mGy/day) for 1 year, while irradiation at a higher dose-rate of 12.5 R/day (125 mGy/h) decreased the number of buds formed down to 0.5% of the control [18]. Prior to the decrease in the number of buds formed, changes in apical meristematic tissues were observed to form; specifically necrosis of apical initial cells developed in the meristem at 22 days after the start of irradiation at 3.75 R/day (37.5 mGy/day). In white spruce (*Picea glauca*), the disappearance of apical initial cells was observed even in the lowest dose-rate of 0.5 R/day (5 mGy/ day) after being exposed to a total dose of 52.7 R (527 mGy) from spring through to autumn [19]. In pitch pine, changes in the apical meristem were also observed after 2 months of chronic exposure at 5 R/day (50 mGy/day) [17]. In these injured meristem, regeneration of meristematic tissue was often observed to replace the necrotic apical meristem [17–19].

In the case of reproductive organs, the number of mature seeds in pitch pine decreased down to 10% of the control, after 9 years of irradiation at 3.5 R/day (35 mGy/day) [12]. Additionally, around 50% of the pollen aborted at 7 R/day (70 mGy/day), and germination of pollen decreased to 50% of normal at 3–5 R/day (30–50 mGy/day) [20]. Male flowers tended to shorten in length at 0.4 R/day (4 mGy/day), and such change became significant at 3.7 R/day (37 mGy/day) [20].

8.3 Radiation Injuries to Coniferous Trees Due to the Chernobyl Nuclear Power Plant (CNPP) Accident

In the CNPP accident in the spring of 1986, radioactive plumes contaminated a wide range of surrounding forests rich in subarctic coniferous trees. The results of the examination conducted by the former Soviet Union (USSR) on the radiation effect of forest plants have been summarized in the report by IAEA [7]. It has been shown that radiation injuries, including withering and death, occurred in forest trees contaminated with radionuclides in the Chernobyl exclusion zone (CEZ), particularly in coniferous trees.

The main coniferous trees that constitute forests in this area are Scots pine (*Pinus sylvestris*) and Norway spruce (*Picea abies*), both of which showed clear biological injuries. In "lethal zone" (4 km^2), which suffered the highest degree of contamination, all of these coniferous trees died and were called "red forest" from the color of the dead trees. In areas with lower level of radiation contamination, the coniferous injuries reduced, and in "sublethal zone" (38 km^2), there were some individual tree deaths or deaths of most growth points. Reproductive suppression and morphological changes were observed in the "medium damage zone" (120 km^2), while even the "minor damage zone" showed growth, reproduction and morphological disturbances. In contrast, radiation injury to broad-leaved trees such as white birch (*Betula platyphylla*) that were mixed with these coniferous trees was relatively small, and even in the "red forest" where all the coniferous trees had died, only partial injuries were observed.

Radiation injuries, such as death of coniferous trees, were observed within 2–3 weeks after the CNP accident and progressed from the spring to the summer in that year. For such injuries, the contribution of radiation exposure in the short term in the early stages of the accident was large, and it was thought that the majority was caused by β-particles from radionuclides directly adhering to the trees. According to the dose estimation of radiation exposure of trees, the reproductive capacity in Scots pine was affected by receiving a total dose of 2000 mGy or more at subacute exposure; growth retardation and morphological damage occurred at 1000 mGy or more; cytogenetic damage was estimated to have been detectable at 500 mGy or more; and there were no visible damages until 5 years after the accident at doses under 100 mGy [21]. In Norway spruce, the radiation sensitivity was higher than that of Scots pine, and the absorbed dose to cause needle, bud, and stem malformations was estimated to be 700–1000 mGy.

In contrast, although the number of studies on the longer-term chronic radiation effect is limited, it has been reported that Scots pine planted after the cutting of the dead trees in the "red forest" showed delayed morphological changes several years later, which has led to abnormal branching of the trunk (disappearance of apical dominance) [22]. In the area where such a morphological change has occurred in 50% of individual trees, the radiation dose-rate of trees was estimated to be 0.96 mGy/day [22].

8.4   Changes in Forests After the Fukushima Daiichi Nuclear Power Plant (FNPP) Accident

Around FNPP, there were a number of warm temperate forests on the land where radionuclides flowed after the accident in March 2011 were deposited. In the forests of this region, coniferous trees such as Japanese red pine (*Pinus densiflora*), Japanese cedar (*Cryptomeria japonica*), Japanese cypress (*Chamaecyparis obtusa*), and Japanese fir (*Abies firma*) are mixed with broad-leaved trees like Konara oak (*Quercus serrata*). These coniferous trees have been mainly examined in forests in the ex-evacuation zone [1]. During the first investigation conducted within the ex-evacuation zone of the FNPP accident in November 2011, around 8 months after the accident, external signs of radiation injuries such as yellowing of the leaves of trees and morphological abnormalities were not observed, even in the most contaminated forest around 3 km west of FNPP [1, 23]. This indicates that large-scale radiation injury in coniferous trees due to subacute radiation exposure did not occur after the FNPP accident, unlike after the CNPP accident. However, to investigate the mid- and long-term effect on trees by radiation, examination within the ex-evacuation zone was continued.

In an examination in January 2015, around 4 years after the accident, the tree morphology in the Japanese fir population naturally occurring in the forest was observed [24]. In each of the three experimental sites within the ex-evacuation zone and one control site that was quite distant, plots of 800–1200 m^2 were established, and all young fir trees (height of 40 cm–5 m, around 100–200 trees in each plot) were observed (Fig. 8.1). In the Japanese fir population in the sites with particularly high ambient dose-rate in the ex-evacuation zone, a significant increase in the frequency of morphological changes was observed compared to the population in the control site with low ambient dose-rate (Fig. 8.2). Moreover, it was found that the frequency was higher depending on the ambient dose-rate in each site. Whereas Japanese fir trees usually have one trunk extending vertically, trees with morphological changes were characterized by branching defects caused by deletion of the leader shoot. When the position of branching defect was identified for each individual tree, a significant increase in the frequency of branching defect became apparent in the part that elongated after the accident between 2012 and 2013, compared to before the accident in 2010 (Fig. 8.3). Taking into account that, as shown in past γ-irradiation experiments and the CNPP accident, coniferous trees like Japanese fir have high radiation sensitivity, these results suggest the possibility that radiation contributed to morphological changes in the Japanese fir trees in area near FNPP.

Similar changes to coniferous trees have been reported in Japanese red pine as well [26]. As a result of investigating the morphology of young Japanese red pine in eight sites within Fukushima Prefecture (including the ex-evacuation zone) between 2014 and 2016, disappearance of apical dominance (i.e., branching of trunk) was observed as a typical morphological change. The incidence rate of morphological changes correlated with the dose-rate of radiation that the tree population was

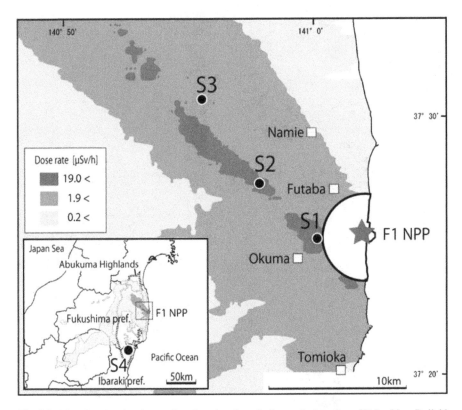

Fig. 8.1 Map showing the observation sites. A red star indicates the location of Fukushima Daiichi Nuclear Power Plant. The base map was modified from the Digital Japan Portal Web Site (Geospatial Information Authority of Japan; GSI). Airborne monitoring results on 2013.11.19; the "Extension site of distribution map of radiation dose, etc.,/Digital Japan" [25] was used as the source of the distribution of radiation dose

exposed to, and the morphological changes manifested within 4 years from the start of exposure. These were similar to the morphological changes observed in the Scots pine within the ex-evacuation zone following the CNPP accident [22], which suggested that radiation is the cause of morphological changes.

8.5 Perspectives for Future Study

The investigations on the coniferous trees within the ex-evacuation zone of the FNPP accident did not show the kind of large-scale radiation injuries observed after the subacute radiation exposure at the initial stages following the CNPP accident. On the other hand, the fact that morphological changes were observed more frequently in shoots of Japanese fir and Japanese red pine near FNPP suggests the influence of chronic exposure. However, it is important to note that the observed

Fig. 8.2 Relative
frequency of main axis
defects in Japanese fir trees
from different sites. The
defects were counted by
the observation of five
annual whorls from the top
of the trees. The figure in
parenthesis indicates
ambient dose-rate at each
observation site.
Bonferroni-corrected
p-values are presented
using Chi-square tests with
df = 1

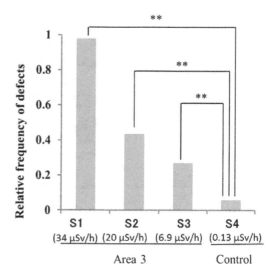

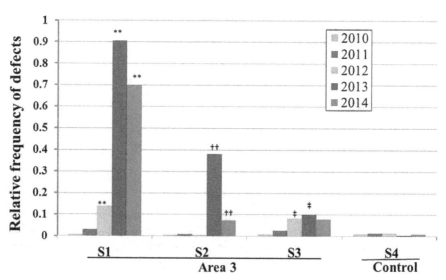

Fig. 8.3 Relative frequency of deleted leader shoot in the annual whorls of the main axis in
Japanese fir trees from different sites. The deletions were counted by the observation of five annual
whorls from the top of the trees. Years indicate the time that the whorl branches sprouted out in the
spring. Different letters indicate statistically significant differences compared to the whorls of
2010 in each site

morphological changes in coniferous trees are not necessarily radiation-specific,
and similar changes may also be caused by environmental and physiological factors. Therefore, in order to clarify the relationship between radionuclide contamination and morphological changes, it is necessary to evaluate accurately the difference
in exposed radiation dose of coniferous trees and various environmental factors in

Fig. 8.4 Chronic irradiation experiment on Japanese fir trees in the gamma field at the National Agriculture and Food Research Organization, Japan

each area of investigation, to specifically elucidate the process by which the morphological changes developed.

Additionally, in order to verify the morphological changes in coniferous trees due to radiation exposure, it is important to make comparisons with data from irradiation experiments that used experimental facilities like gamma fields. Existing data from irradiation experiments conducted on coniferous trees indicate that radiation sensitivity differs depending on the coniferous species and depending on the injuries such as lethality, growth inhibition, effects on reproductive function, etc. Therefore, irradiation experiments on the specific coniferous species (Japanese fir and Japanese red pine) should be considered, under conditions similar to their growing environments, to investigate if the same morphological change as observed near FNPP occurs (Fig. 8.4).

To reproduce the radiation exposure situation in the verification experiment at the irradiation facility, it is also important to estimate as accurately as possible the radiation dose that the coniferous trees were actually exposed to in the investigated area. By estimating the radiation dose of the coniferous trees in the investigated area and comparing it with the irradiation dose at the radiation facility that causes morphological changes in the coniferous trees, it will become possible to elucidate the causal relationship between the incidence of morphological changes in the investigated area and radiation exposure. For this reason, it is an important challenge to create a dose evaluation model to reproduce complicated exposures that coniferous tree tissues receive from various radiation sources in the environment and then reconstruct the change in the exposed dose of coniferous trees after the FNPP accident based on actual measurements of the radionuclide concentration in the trees and their environment.

8.6　Remarks

The contents concerning morphological abnormalities posted here are based on "Morphological defects in native Japanese fir trees around the Fukushima Daiichi Nuclear Power Plant" by Y. Watanabe et al., licensed under Creative Commons.

References

1. Ministry of Environment Results of measurements after monitoring wild animals and plants. https://www.env.go.jp/jishin/monitoring/results_r-wl.html. Accessed 20 Mar 2018
2. Sparrow AH (1966) Research uses of the gamma field and related radiation facilities at Brookhaven national laboratory. Radiat Bot 6:377–405
3. Kawara K (1963) Introduction of a gamma field in Japan. Radiat Bot 3:175–177
4. Sparrow AH, Schairer LA, Woodwell GM (1965) Tolerance of *Pinus rigida* trees to a ten-year exposure to chronic gamma irradiation from cobalt-60. Radiat Bot 5:7–22
5. Dugle JR (1986) Growth and morphology in Balsam fir: effects of gamma radiation. Can J Bot 64:144–192
6. United Nations Scientific Committee on the Effects of Atomic Radiation (1996) Sources and effects of ionizing radiation; report to the general assembly. United Nations, New York
7. International Atomic Energy Agency (2006) Environmental consequences of the Chernobyl accident and their remediation: twenty years of experience, report of the Chernobyl Forum Expert Group 'Environment'. IAEA, Vienna
8. Sparrow RC, Sparrow AH (1965) Relative radiosensitivities of woody and herbaceous spermatophytes. Science 147:1449–1451
9. Sparrow AH, Rogers AF, Schwemmer SS (1968) Radiosensitivity studies with woody plants. I. Acute gamma irradiation survival data for 28 species and predictions for 190 species. Radiat Bot 8:149–174
10. Sparrow AH, Schwemmer SS, Klug EE et al (1970) Radiosensitivity studies with woody plants. II. Survival data for 13 species irradiated chronically for up to 8 years. Radiat Res 44:154–177
11. Sparrow AH, Schwemmer SS, Klug EE et al (1970) Woody plants: changes in survival in response to long-term (8 years) chronic gamma irradiation. Science 169:1082–1084
12. Ohba K (1964) Studies on radiosensitivity and induction of somatic mutations in forest trees. Gamma Field Symp 3:111–141
13. Murray BG (1998) Nuclear DNA amounts in gymnosperms. Ann Bot 82:3–15
14. Baetcke KP, Sparrow AH, Nauman CH et al (1967) The relationship of DNA content to nuclear and chromosome volumes and to radiosensitivity (LD50). Proc Natl Acad Sci USA 58:533–540
15. Sparrow AH, Woodwell GM (1962) Prediction of the sensitivity of plants to chronic gamma irradiation. Radiat Bot 2:9–26
16. Murai M, Ohba K (1966) Study on radiosensitivity of forest trees (III) influence of gamma ray irradiation on the growth of *Larix Kaempferi* larch and cell division in leaves. J Jpn For Soc 48:62–68
17. Mergen F, Thielges BA (1966) Effects of chronic exposures to Co60 radiation on *Pinus rigida* seedlings. Radiat Bot 6:203–210
18. Miksche JP, Sparrow AH, Rogers AF (1962) The effects of chronic gamma irradiation on the apical meristem and bud formation of Taxus media. Radiat Bot 2:125–129
19. Cecich RA, Miksche JP (1970) The response of white spruce (*Picea glauca* (Moench) Voss) shoot apices to exposures of chronic gamma radiation. Radiat Bot 10:457–467

20. Mergen F, Johansen TS (1963) Effect of ionizing radiation on microsporogenesis in *Pinus Rigida* mill. Radiat Bot 3:321–331
21. Arkhipov NP, Kuchma ND, Askbrant S et al (1994) Acute and long-term effects of irradiation on pine (Pinus silvestris) stands post-Chernobyl. Sci Total Environ 157:383–386
22. Yoschenko VI, Kashparov VA, Melnychuk MD et al (2011) Chronic irradiation of scots pine trees (*Pinus Sylvestris*) in the Chernobyl exclusion zone: dosimetry and radiobiological effects. Health Phys 101:393–408
23. Watanabe Y, Ichikawa S, Kubota M et al (2012) Effects of radionuclide contamination on forest trees in the exclusion zone around the Fukushima Daiichi Nuclear Power Plant. In: Proceedings of the international symposium on environmental monitoring and dose estimation of residents after accident of TEPCO's Fukushima Daiichi Nuclear Power Stations, pp 2–14
24. Watanabe Y, Ichikawa S, Kubota M et al (2015) Morphological defects in native Japanese fir trees around the Fukushima Daiichi Nuclear Power Plant. Sci Rep 5:13232
25. Ministry of Education, Culture, Sports, Science and Technology Radiation dose distribution map, http://ramap.jaea.go.jp/map/. Accessed 20 Mar 2018
26. Yoschenko V, Nanba K, Yoshida S et al (2016) Morphological abnormalities in Japanese red pine (*Pinus densiflora*) at the territories contaminated as a result of the accident at Fukushima Dai-Ichi Nuclear Power Plant. J Environ Radioact 165:60–67

Part III
Impact of Radiation on Livestock

9

Female Fertilities of Domestic Animals: Impact of Low-Dose-Rate Radiation

Yasuyuki Abe, Hideaki Yamashiro, Yasushi Kino, Toshinori Oikawa,
Masatoshi Suzuki, Yusuke Urushihara, Yoshikazu Kuwahara,
Motoko Morimoto, Jin Kobayashi, Tsutomu Sekine, Tomokazu Fukuda,
Emiko Isogai, and Manabu Fukumoto

Abstract In order to understand the effect of chronic exposure to low-dose-rate
radiation on female fertilities and the risk of heritable genetic effects on domestic
animals after the Fukushima Daiichi Nuclear Power Plant (FNPP) accident, we
assessed the developmental ability of oocytes, and examined the histological
characters and structure of ovaries. We showed that potentially viable oocytes could
be collected from the ovary of abandoned cattle in the ex-evacuation zone set within
a 20-km radius from FNPP, resulting in production of the morphologically normal
calves following in vitro culture and embryo transfer. The proliferation of cattle
ovarian granulosa cells was confirmed by expression of Ki-67. Apoptosis of oocytes
and granulosa cells was few determined by the TUNEL assay. In addition, porcine
and inobuta oocytes had the abilities of in vitro maturation. These results suggest
that chronic radiation exposure associated with the FNPP accident may have little
effect, if any, on the female fertilities of domestic animals.

Y. Abe (✉)
Faculty of Life and Environmental Sciences, Prefectural University of Hiroshima,
Shobara, Hiroshima, Japan
e-mail: abe@pu-hiroshima.ac.jp

H. Yamashiro
Graduate School of Science and Technology, Niigata University, Niigata, Japan

Y. Kino
Department of Chemistry, Tohoku University, Sendai, Japan

T. Oikawa
Miyagi Prefectural Livestock Experiment Station, Oosaki, Miyagi, Japan

M. Suzuki · Y. Kuwahara
Institute of Development, Aging and Cancer, Tohoku University, Sendai, Japan

Keywords Fukushima Daiichi Nuclear Power Plant · Radioactive cesium · Cattle · Ovary · Oocyte · Fertility

9.1   Introduction

Female reproductive organs such as the ovary, oviduct and uterus are essential organs to produce the next generation in mammals. Oocytes, female germ cells, are contained in the ovary. Oogenesis, namely, oocyte production begins in the fetal ovary. Oogonia proliferate via mitosis ultimately become oocytes via meiosis, and are arrested in the first meiotic prophase during fetal development [9]. The ovaries at the time of birth contain a large amount of oocytes, and oogenesis in the adult does not increase the number of germ cells. In the ovary, oocytes form the ovarian follicles along with somatic cells such as granulosa cells which support the growth of oocytes. A primordial follicle, consisting of an oocyte surrounded by a single layer of follicular cells (pregranulosa cells), is the most immature stage of an ovarian follicle's development and can remain in this state for many years. After puberty (sexual maturation), some primordial follicles resume development into the primary, secondary and antral follicles with proliferation of granulosa cells and an increase in the oocyte diameter, and finally ovulate following development to the fully mature state. Damage to oocytes and follicular cells may lead to apoptosis, mutants and chromosomal changes, resulting in infertility and hereditary diseases.

Due to the nuclear accident caused by the Great East Japan Earthquake on 11 March 2011, a large amount of radioactive substances were emitted from Fukushima Daiichi Nuclear Power Plant (FNPP) [6, 15]. Although there are great concerns about the effect of emitted radioactive substances on female germ cells and fetuses in humans, radiobiological data on those risks led by chronic radiation exposure to low-dose-rate radiation are limited. In radiation therapy, previous studies have

Y. Urushihara
Graduate School of Medicine, Tohoku University, Sendai, Japan

M. Morimoto · J. Kobayashi
School of Food, Agricultural, and Environmental Sciences, Miyagi University, Sendai, Japan

T. Sekine
Institute for Excellence in Higher Education, Tohoku University, Sendai, Japan

T. Fukuda
Graduate School of Science and Engineering, Iwate University, Morioka, Japan

E. Isogai
Graduate School of Agricultural Sciences, Tohoku University, Sendai, Japan

M. Fukumoto
Institute of Development, Aging and Cancer, Tohoku University, Sendai, Japan

School of Medicine, Tokyo Medical University, Tokyo, Japan

shown that the human oocytes are sensitive to radiation and that the estimated median lethal dose (LD_{50}) is less than 2 Gy [12]. On the other hand, Adriaens et al. [2] reviewed that congenital anomalies have been observed after exposure to high doses (1–5 Gy) in mice, but prudence is required to extrapolate these data to humans. However, it is difficult to obtain a clear answer for risks of low-dose radiation only by epidemiological studies because of the diversity of examined population and the requirement of a long-term observation. Furthermore, careful and accurate analyses are required. The domestic animals abandoned in the ex-evacuation zone set inside a 20-km radius surrounding FNPP (abandoned animals) were the invaluable research samples for studying the effect of chronic radiation to low-dose-rate exposure. In addition, it is important to preserve the samples for the advanced analysis in the future [11].

In order to elucidate the effect of chronic radiation exposure on female fertilities and the risk of heritable genetic effects in domestic animals after the FNPP accident, we assessed the developmental ability of oocytes and performed histological examination in ovaries.

9.2   Methods

9.2.1   Production of the Next Generation Derived from Abandoned Animals

The present study was approved by the Ethics Committee of Animal Experiments, Tohoku University. Cattle ovaries of Japanese Black and Holstein were collected in Kawauchi Village located 15 km southwest, Naraha Town located 17 km south, and Tomioka Town located 10 km south of FNPP between August 31, 2011 and September 12, 2012 (from 6 months to 1.5 years after the accident). The collected ovaries were washed and stored at approximately 20 °C in physiological saline containing antibiotics and transported to the laboratory within 7 h. The bovine embryos were in vitro produced and cryopreserved as described [1] with the following modifications. The bovine cumulus-oocyte complexes (COCs) were aspirated from visible ovarian follicles by a 10-ml syringe with an 18-gage needle and were put into the maturation medium consisting of TCM199 supplemented with 10% fetal bovine serum (FBS), follicle-stimulating hormone, and antibiotics. After the culture for 22–23 h, the matured COCs were co-incubated with frozen-thawed spermatozoa in the fertilization medium consisting of Brackett and Oliphant medium with 3 mg/ml bovine serum albumin (BSA) and 2.5 mM theophylline. Those spermatozoa were commercially available and derived from a non-radiocontaminated bull semen of Japanese Black cattle (Livestock Improvement Association of Japan, Tokyo, Japan). After insemination, cumulus cells were removed from presumptive zygotes by putting them in and out using a fine-bore pipette and then cultured for in vitro development to the blastocyst stage in a modified synthetic oviduct fluid medium. After examination, the blastocysts obtained at days 7 to 9 post culture were cryopreserved by a vitrification method. The blastocysts were exposed first for 7 min to solution A, which was composed of 10% (v/v) ethylene glycol, 4.5% (w/v) Ficoll-70 and 0.075 M sucrose in modified phosphate-buffered saline (PB1) [13]; then for 2 min to solution B, which was composed of 20% (v/v) ethylene glycol, 9.0% (w/v) Ficoll-70 and 0.15 M sucrose in PB1; and finally for 1 min to solution C, which was composed of 40% (v/v) ethylene glycol, 18% (w/v) Ficoll-70 and 0.3 M sucrose in PB1 at room temperature. One to two blastocysts were then transferred onto a nylon-mesh holder with a minimum amount of medium, and the holder was directly plunged into liquid nitrogen (LN_2) and stored until use for embryo transfer. Exposing to solution C and plunging into LN_2 were performed within 1 min. To warm the frozen blastocysts, the holder with blastocysts was soaked in PB1 containing 0.5 M sucrose at 37°C. To remove the cryoprotectants, the blastocysts recovered from the holder were exposed to PB1 with a sequential series of 0.5, 0.25 and 0.125 M sucrose for 1 min in each solution at 37°C and finally transferred into PB1 for 5 min at 37°C. Eight Japanese Black blastocysts with normal morphology were transferred nonsurgically to synchronized recipient cows (one blastocyst per animal).

We could also collect ovary samples from 12 pigs and 2 inobutas (hybrids of a wild boar and a pig) between 18 January 18 and February 28, 2012 (10–11 months from the accident). Different from cattle ovaries, their ovaries were transported to the laboratory at 37°C. In vitro maturation (IVM) was carried out for 44 h on COCs, instead of 22–23 h in cattle. The presumptive matured oocytes were vitrified by the similar manner to the above for cattle.

9.2.2 Morphology, Cell Proliferation and Apoptosis Assay of the Ovary

After aspiration of COCs, the ovary was fixed in 10% buffered formalin and was embedded in paraffin and sectioned. General morphology was assessed by hematoxylin and eosin staining, granulosa and theca cell proliferation index was by Ki-67 immunostaining [8], and apoptosis of granulosa cells was by terminal deoxynucleotidyl transferase biotin-dUTP nick end labeling (TUNEL) assay [3].

9.2.3 Measurements of Radioactivity

Radioactivity of the female reproductive organs (ovaries, oviducts, and uteri) and peripheral blood was determined by γ-ray spectrometry using three high-purity germanium detectors (ORTEC, TN, USA) described in the previous report [14]. The measurement time took 3,600–200,000 s depending on the radioactivity of the sample. The detection efficiency was determined by measuring mixed sources of cesium-134 (^{134}Cs) and cesium-137 (^{137}Cs). An aliquot (200 ml) of the mixing source was diluted with an appropriate amount of water, and a superabsorbent polymer was added to the mixture to make a gel standard source, which was used as a simulated sample mimicking biological samples. Several types of gel sources were prepared to cover the weight range from 0.5 to 130 g. The detection efficiency of the liquid samples such as peripheral blood was determined using aqueous solutions of ^{134}Cs and ^{137}Cs.

9.3   Results

9.3.1   Viability and Fertility of Oocytes in Abandoned Animals

A total of 714 morphologically viable COCs were collected from 42 adult cattle, among which some COCs had partly dispersed cumulus cells or heterogeneous ooplasm. Figure 9.1 shows the morphological appearances of excised female reproductive organs (1A), ovaries (1B), COCs collected from ovaries (1C), COCs in vitro matured for 23 h (1D), and produced embryos (1E) following in vitro fertilization (IVF) and culture (IVC) for 8 days. No remarkable abnormalities were observed in the collected organs. After culture for IVM, cumulus expansion with cell proliferation, which is known as an index of oocyte maturation, was observed in almost all COCs. As a result of IVF and IVC, the rates of cleavage and development to blastocyst stage were 54.9% and 10.5%, respectively. Finally, 41 blastocysts were yielded and then were cryopreserved. After the transfer of eight blastocysts to eight synchronized recipient cows, three calves were born (Fig. 9.1F, Table 9.1). A total of 436 porcine and 64 inobuta COCs were collected from 12 and 2 adult animals, respectively, and the COCs exhibited the cumulus expansion after IVM. The rate of presumptive maturation was 85.1% (371/436) for pigs and 100% (64/64) for inobutas. These presumptive matured oocytes were cryopreserved for further analyses.

9.3.2   Cell Proliferation and Apoptosis in Cattle Ovaries

The histological examination (Fig. 9.2) revealed that none of the ovaries derived from 12 adult abandoned cattle (2A and B) and 2 infants (2C and D) exhibited structural organ damage, as shown in Fig. 9.2. Although the abandoned cattle can be recognized individually by the ear tag, the age of the cattle without the ear tag was estimated from height and body size. We classified a cattle as infant if it was younger than 6 months old. Those ovaries had primordial, primary, secondary and antral follicles, which were morphologically intact. Furthermore, Ki-67 immunoreactivity was detectable in many granulosa cells of the primary and secondary follicles (Fig. 9.3). The follicle with TUNEL-positive apoptotic cells was observed in only 1 follicle from a cattle, but the follicular cells were TUNEL-negative in a total of 97 follicles derived from 11 cattle.

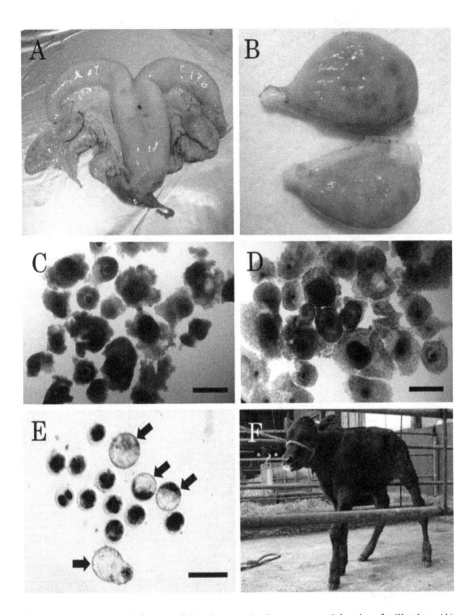

Fig. 9.1 Morphological figures of female reproductive organs and in vitro fertilization. (**A**) Excised female reproductive organs. (**B**) Some follicles are visible through the ovarian surface. (**C**) Cumulus-oocyte complexes (COCs) collected from the ovaries. (**D**) In vitro matured COCs exhibiting cumulus expansion, that is, an index of oocyte maturation. Arrows indicate the produced blastocysts. Bar, 500 μm. (**E**) Developed embryos following in vitro fertilization and cultured for 8 days. (**F**) A delivered calf following embryo transfer

Table 9.1 Embryo transfer of produced cattle blastocysts

Doner				
Date of collection	Habitat	^{137}Cs (Bq/kg)[a]	No. of embryos transferred	No. of calves born
24 January 2012	Tomioka	26 ± 1	1	1
24 January 2012	Tomioka	30 ± 1	3	1
24 January 2012	Tomioka	33 ± 1	1	1
6 December 2012	Kawauchi	20 ± 1	1	0
24 January 2012	Tomioka	28 ± 1	2	0

[a] ^{137}Cs activity concentration in peripheral blood

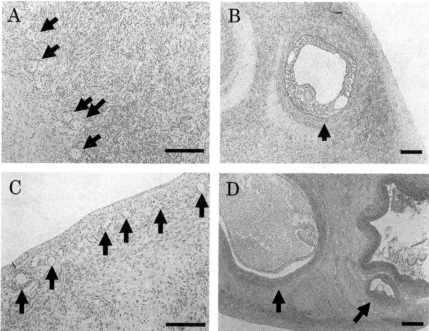

Fig. 9.2 Histology of the ovary derived from (**A, B**) an adult cattle and (**C, D**) an infant estimated to be 3 months old. These ovaries have growing follicles from primordial to antral stage. Arrows indicate the follicles. Bar, 100 μm

9.3.3 Radioactivity Concentration of ^{134}Cs and ^{137}Cs in Female Cattle Reproductive Organs

The reproductive organs such as ovaries, oviducts and uterus were derived from 18 adult cattle, 5 infants and 2 fetuses. Concentrations of ^{134}Cs and ^{137}Cs in female reproductive organs are shown in Tables 9.2 and 9.3, respectively. After decay correction to the day of major release, March 15, 2011, radioactivity concentration of

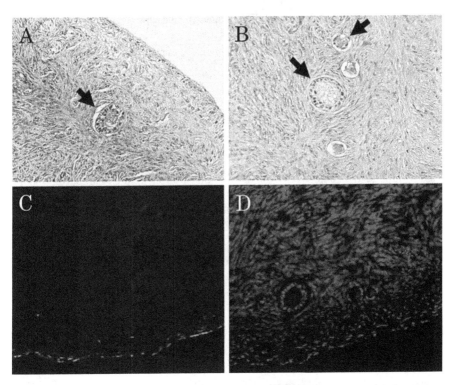

Fig. 9.3 Immunohistochemistry of (**A**, **B**) Ki-67, an proliferated cell marker; (**C**) TUNEL, an apoptotic cell marker; and (**D**) propidium iodide, counterstain, of ovaries derived from adult cattle. Follicles have the proliferated and non-apoptotic granulosa cells. Arrows indicate the follicles with proliferated granulosa cells

Table 9.2 Radioactivity concentration of ^{134}Cs in female reproductive organs and peripheral blood of cattle (Bq/kg)

Origin	Mean± SEM of ^{134}Cs concentration[a] (no. of samples)			
	Ovary	Oviduct	Uterus	Blood
Adult	136 ± 19 (11)	65 ± 12 (10)	98 ± 18 (18)	28 ± 3 (58)
Infant	135 ± 8 (2)	26 ± 11 (2)	140 ± 58 (5)	31 ± 6 (8)
Fetus	48 ± 15 (2)	30 ± 19 (2)	62 ± 24 (2)	ND[b] (0)

[a]Decay correction was made to the day of major release of radionuclides, March 15, 2011
[b]No data

^{134}Cs was almost equal to that of ^{137}Cs in all organs regardless of the age. The radioactivity concentration levels of ^{137}Cs in the ovary of adult cattle (135 ± 19 Bq/kg; range 11–256 Bq/kg) and infants (137 ± 12Bq/kg; range 124–149 Bq/kg) were higher than those levels in the ovaries of fetuses (46 ± 15 Bq/kg; range 31–61 Bq/kg). Also in oviducts and uterus, the levels in adult cattle and infants were lower than those in fetuses. On the other hand, ^{137}Cs concentration in ovaries was the

Table 9.3 Radioactivity concentration of ^{137}Cs in female reproductive organs and peripheral blood of cattle (Bq/kg)

Origin	Mean ± SEM of ^{137}Cs concentration[a] (no. of samples)			
	Ovary	Oviduct	Uterus	Blood
Adult	135 ± 19 (11)	66 ± 12 (10)	99 ± 19 (18)	29 ± 3 (58)
Infant	137 ± 12 (2)	25 ± 10 (2)	142 ± 59 (5)	32 ± 6 (8)
Fetus	46 ± 15 (2)	32 ± 19 (2)	63 ± 25 (2)	ND[b] (0)

[a]Decay correction was made to the day of major release of radionuclides, March 15, 2011
[b]No data

highest among the female reproductive organs examined and was approximately 2.3-fold greater than that in oviducts.

The peripheral blood samples were derived from 58 adult cattle and 8 infants. As with the reproductive organs, radioactivity concentration of [134]Cs in peripheral blood was almost equal to that of [137]Cs. The levels of [137]C concentration in adult cattle (29 ± 3 Bq/kg; range 4–96 Bq/kg) were similar to those in infants (32 ± 6 Bq/kg; range 12–62 Bq/kg). The levels in peripheral blood were more than 3-folds smaller than those in ovaries and uterus regardless of the age.

9.4   Discussion

In the present study, we showed that potentially viable oocytes were obtainable from the ovaries of abandoned cattle, resulting in production of offspring. The number of collected oocytes per animal (17.0) was equal or more compared with that of those collected in a slaughter house (6–10). On the other hand, the rates of cleavage and development to blastocyst stage (54.9% and 10.5%, respectively) were lower than those in the ordinary experiments with the samples from the slaughter house (70–80% and 20–40%, respectively). In this study, the collection of organs was carried out in the unsuitable condition for oocytes: an open field with harsh temperature, euthanasia by anesthetic agents that may have an adverse effect on oocytes, and a long time between sampling and the start of IVM. Additionally, COC morphology is involved in the determination of in vitro developmental competence of immature oocyte; therefore the COCs with compact cumulus cells and homogeneous ooplasm are selected for in vitro embryo production [5]. In the present study, we performed IVM for COCs even with partly dispersed cumulus cells or heterogeneous ooplasm because the primary objective was to have offspring production. Although the blastocyst rate was low for the above reasons, the collected oocytes had the developmental competence to the blastocyst stage regardless the habitat and ^{137}Cs activity concentration in peripheral blood of their originating animals (1–33 Bq/kg). In addition, the delivery rate following cryopreservation and transfer of produced blastocysts (37.5%, 3/8) was similar to the rate of non-radiocontaminated ones in our facility (approximately 40%), and the born calves were macroscopically normal.

Although several studies with laboratory animals on the genetic effect of ioniz-ing radiation showed that radiosensitivity of oocytes varies widely according to the follicle/oocyte stage and the species [2], the effect on large animals such as cattle remains unknown. In the present study, histological examination of cattle ovaries showed that there was not much difference in the morphology and the average num-ber of the follicles between ovaries collected from cattle abandoned and those col-lected at the slaughterhouse. The intact proliferation in cattle ovarian granulosa cells was confirmed by expression of Ki-67, and quite few apoptotic oocytes and granu-losa cells were observed by the TUNEL assay. Granulosa cells produce estradiol-17β and provide various low-molecular-weight substrates (e.g., amino acids and nucleotides) to the oocytes during folliculogenesis [7, 10]. Those compounds are utilized to synthesize important macromolecules such as proteins and nucleic acids. Estradiol-17β acts as a growth hormone for tissues of the reproductive organs and appears necessary to maintain oocytes in the ovary. Apoptosis of granulosa cells is associated with induction of follicular atresia in the mammalian ovary. Moreover, the dysfunction of granulosa cells leads to the arrest of oocyte growth and follicular development, resulting in infertility. Our data in histological examination of ovaries and the fact that some cattle captured in the ex-evacuation zone were pregnant sug-gest that cattle ovaries appear to have the potentials of endocrine function and ovu-lation, and that the folliculogenesis was performed normally in abandoned cattle.

Regarding radioactive Cs concentration, the level in the female reproductive organs of fetuses was lower than those in adult cattle and infants, although the level in other organs of fetus and infants was higher than that of the corresponding maternal organ, respectively [4]. Since active angiogenesis occurs during the late follicular development after puberty and during gestation, blood vessels in the fetal reproductive organs are premature, and, therefore, the amount of transported radioactive Cs into the fetus may be limited. The levels of radioactive Cs concentration in the oviducts were lower than those in the ovary and uterus, which may be because the oviduct with luminal structure is composed of less parenchymal cells than the ovary and uterus. As well as cattle, we confirmed that porcine and inobuta ovaries maintained competence for oocyte maturation in vitro without mor-phologically abnormal findings.

In conclusion, no adverse radiation-induced effects were observed in the female reproductive organs derived from the domestic animals exposed to chronic radiation for 379 days in the ex-evacuation zone. The oocyte maintained the developmental competence to potentially viable blastocysts, and no hereditary abnormalities were observed in the yield offspring. In addition, we succeeded in cryopreserving the oocytes of the corresponding animals. Unfortunately, even if such accidents or problems as the FNPP accident occur in the future, we hope that our data and cryo-preserved materials will be of some help.

Acknowledgments We would like to thank the Iwaki Livestock Hygiene Service Centre in Fukushima Prefecture, all people associated with the livestock production field in Fukushima. This study was supported by the Emergency Budget for the Reconstruction of Northeastern Japan, MEXT, Japan; Discretionary Expense of the President of Tohoku University; and Nippon Life Insurance Foundation.

References

1. Abe Y, Hara K, Matsumoto H et al (2005) Feasibility of a nylon-mesh holder for vitrification of bovine germinal vesicle oocytes in subsequent production of viable blastocysts. Biol Reprod 72(6):1416–1420
2. Adriaens I, Smitz J, Jacquet P (2009) The current knowledge on radiosensitivity of ovarian follicle development stages. Hum Reprod Update 15(3):359–377
3. Brooks K, Spencer TE (2015) Biological roles of interferon tau (IFNT) and type I IFN receptors in elongation of the ovine conceptus. Biol Reprod 92(2):47. 1–10
4. Fukuda T, Kino Y, Abe Y et al (2013) Distribution of artificial radionuclides in abandoned cattle in the evacuation zone of the Fukushima Daiichi nuclear power plant. PLoS One 8(1):e54312
5. Hazeleger NL, Hill DJ, Stubbing RB, Walton JS (1995) Relationship of morphology and follicular fluid environment of bovine oocytes to their developmental potential in vitro. Theriogenology 43(2):509–522
6. Kinoshita N, Sueki K, Sasa K et al (2011) Assessment of individual radionuclide distributions from the Fukushima nuclear accident covering central-east Japan. Proc Natl Acad Sci USA 108(49):19526–19529
7. McLaughlin EA, McIver SC (2009) Awakening the oocyte: controlling primordial follicle development. Reproduction 137(1):1–11
8. Salvetti NR, Stangaferro ML, Palomar MM et al (2010) Cell proliferation and survival mechanisms underlying the abnormal persistence of follicular cysts in bovines with cystic ovarian disease induced by ACTH. Anim Reprod Sci 122(1–2):98–110
9. Schillo KK (2008) Reproductive physiology of mammals: from farm to field and beyond. Delmar, New York
10. Shimizu (2016) Molecular and cellular mechanisms for the regulation of ovarian follicular function in cows. J Reprod Dev 62(4):323–329
11. Takahashi S, Inoue K, Suzuki M et al (2015) A comprehensive dose evaluation project concerning animals affected by the Fukushima Daiichi Nuclear Power Plant accident: its set-up and progress. J Radiat Res 56(Suppl 1):i36–i41
12. Wallace WH (2011) Oncofertility and preservation of reproductive capacity in children and young adults. Cancer 117(10 Suppl):2301–2310
13. Whittingham DG (1974) Embryo banks in the future of developmental genetics. Genetics 78(1):395–402
14. Yamashiro H, Abe Y, Fukuda T et al (2013) Effects of radioactive caesium on bull testes after the Fukushima nuclear plant accident. Sci Rep 3:2850
15. Zheng J, Tagami K, Watanabe Y et al (2012) Isotopic evidence of plutonium release into the environment from the Fukushima DNPP accident. Sci Rep 2:304

10

FNPP Accident and the Transgenerational Impact on a Calf

Banri Suzuki, Shigefumi Tanaka, Kohichi Nishikawa, Chikako Yoshida,
Takahisa Yamada, Yasuyuki Abe, Tomokazu Fukuda, Jin Kobayashi,
Gohei Hayashi, Masatoshi Suzuki, Yusuke Urushihara, Kazuma Koarai,
Yasushi Kino, Tsutomu Sekine, Atsushi Takahasi, Toshihiro Shimizu,
Hisashi Shinoda, Kazuki Saito, Emiko Isogai, Koh Kawasumi,
Satoshi Sugimura, Hideaki Yamashiro, and Manabu Fukumoto

Abstract The Fukushima nuclear power plant (FNPP) accident raised worldwide attention to the health risk of radiation exposure and to its potential transgenerational effects. Here, we analysed transgenerational effects on calf spermatogenesis and on blood plasma metabolome in order to detect alterations associated with paternal exposure to low-dose-rate (LDR) radiation. Sperm was collected from a bull exposed to radiation for 2 years abandoned in the ex-evacuation zone of the FNPP accident (the abandoned bull) and was used for artificial insemination (AI) into a non-radiocontaminated cow. Haematoxylin and eosin stained sections of the testis of a 13-month-old calf revealed spermatogonia, spermatocytes, spermatids, and sperm in normal morphology. Nuclear and acrosomal morphology of sperm

B. Suzuki · S. Tanaka · K. Nishikawa · C. Yoshida · T. Yamada · H. Yamashiro (✉)
Graduate School of Science and Technology, Niigata University, Niigata, Japan
e-mail: hyamashiro@agr.niigata-u.ac.jp

Y. Abe
Department of Life Science, Prefectural University of Hiroshima, Hiroshima, Japan

T. Fukuda
Graduate School of Agricultural Sciences, Iwate University, Morioka, Japan

J. Kobayashi
School of Food, Agricultural, and Environmental Sciences, Miyagi University, Sendai, Japan

G. Hayashi · M. Suzuki
Institute of Development, Aging and Cancer, Tohoku University, Sendai, Japan

was generally normal. Metabolomic profiling of plasma using capillary electrophoresis–mass spectrometry resulted in 104 peaks of candidate compounds suggestive of paternal exposure. A calf was delivered by AI using sperm from the abandoned bull. Regarding glycolysis, the contents of nucleotide sugars tended to be lower in the delivered calf than in the control calf. Among energy carries, AMP and ATP showed different tendency between non-radiocontaminated and delivered calf. In conclusion, there were no apparent transgenerational effects on both spermatogenesis and blood plasma metabolome in a calf obtained by AI using sperm from the abandoned bull exposed to LDR in the ex-evacuation zone of the FNPP accident for about 2 years.

Keywords Calf · Fukushima Daiichi Nuclear Power Plant accident · Low-dose-rate · Metabolome · Spermatogenesis

10.1   Introduction

Following the Fukushima Daiichi Nuclear Power Plant (FNPP) accident, large amounts of radioactive substances, particularly volatile elements such as radioactive iodine (^{131}I, ^{132}I and ^{133}I), cesium (^{134}Cs, ^{136}Cs and ^{137}Cs), tellurium (^{132}Te), and inert gases (e.g., ^{133}Xe), were released into the environment [1]. We have established an archive system composed of livestock and wild animals in a 20-km radius around FNPP, that is, the ex-evacuation zone of the FNPP accident [2–5]. This system provides critical information for the understanding of radioactive contamination, environmental pollution, biodistribution and metabolism, and the biological effects of internal and external exposure in association with dose evaluation. Recently, we

Y. Urushihara
Graduate School of Medicine, Tohoku University, Sendai, Japan

K. Koarai · Y. Kino · T. Sekine
Graduate School of Science, Tohoku University, Sendai, Japan

A. Takahasi · T. Shimizu · H. Shinoda
Graduate School of Dentistry, Tohoku University, Sendai, Japan

K. Saito · E. Isogai
Graduate School of Agricultural Sciences, Tohoku University, Sendai, Japan

K. Kawasumi
School of Veterinary Medicine, Nippon Veterinary and Life Science University, Tokyo, Japan

S. Sugimura
Institute of Agriculture, Tokyo University of Agriculture and Technology, Tokyo, Japan

M. Fukumoto
Institute of Development, Aging and Cancer, Tohoku University, Sendai, Japan

School of Medicine, Tokyo Medical University, Tokyo, Japan

reported organ-specific deposition of the individual radionuclide in abandoned cattle following the FNPP accident. Radioactive caesium (^{134}Cs and ^{137}Cs) is notably detected in all organs examined [6–8]. Further, we reported that spermatogenesis occurred normally in the reproductive organs of bulls and boars that were abandoned in the ex-evacuation zone, following chronic low-dose-rate (LDR) radiation [9, 10]. However, to date there is no clear evidence whether transgenerational effects of radiation exposure in livestock animals associated with the FNPP accident exist or not.

Here, we analysed transgenerational effects on the calf spermatogenesis, and blood plasma metabolome using capillary electrophoresis–mass spectrometry (CE-TOFMS) with paternal exposure to chronic LDR radiation by staying for 2 years in the ex-evacuation zone.

10.2 Materials and Methods

10.2.1 Ethics

The Japanese government ordered Fukushima Prefecture to euthanize cattle in the ex-evacuation zone on May 12, 2011 to prevent radio-contaminated beef products from entering the human food chain. We collected organs and tissues from the euthanized cattle by the combined unit of veterinary doctors belonging to the Livestock Hygiene Service Center of Fukushima Prefecture and those belonging to the Ministry of Agriculture, Forestry and Fisheries, Japan. This study was approved by the Ethics Committee of Animal Experiments, Niigata University, Japan (Regulation No. 27-83-3).

10.2.2 Animals

We collected sperm form the testis of an euthanized Japanese black bull (the abandoned bull) at Tomioka Town, located 7 km south of FNPP, on February 28, 2013, as described previously [10]. Immediately after collection, the sperm was diluted with a Triladyl freezing extender containing egg yolk (Mini Tube, Germany); freezing protocol was performed as described previously [11]. On sampling date, ambient dose equivalent rate was 1.7 µSv/h (using a NaI (Tl) Scintillation Survey Meter), and the time elapsed since the FNPP accident (March 11, 2011) was almost 2 years. Dose rate of both internal and external exposure to ^{134}Cs and ^{137}Cs was estimated according to a modified method as previously described [3].

10.2.3   Artificial Insemination (AI)

Using the sperm of the abandoned bull, a non-radiocontaminated recipient black
cow underwent AI at the experimental farm of Niigata University. A male Japanese
black calf (the delivered calf) was born on August 29, 2015 (approximately 10-month
pregnancy period) from which the testis, caudae epididymides, sperm, and blood
plasma were sampled on September 27, 2016 (at approximately 13 months old)
after euthanasia. As non-radiocontamination control, blood plasma was also collected from a non-castrated male Japanese black calf at almost the same age as the
delivered calf from Niigata Prefecture, at the same latitude as Fukushima but not
affected by the FNPP accident. The relationship between each cattle and artificial
insemination is shown in Fig. 10.1.

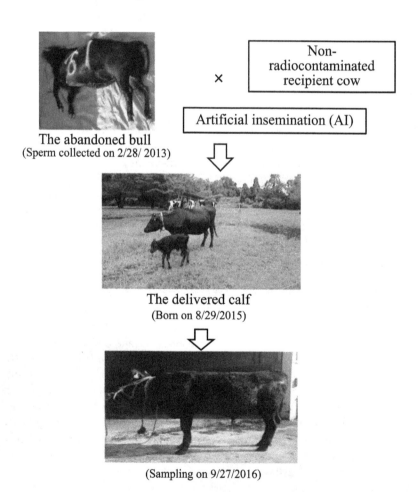

The abandoned bull
(Sperm collected on 2/28/ 2013)

× Non-radiocontaminated recipient cow

Artificial insemination (AI)

The delivered calf
(Born on 8/29/2015)

(Sampling on 9/27/2016)

Fig. 10.1 The delivered calf obtained by artificial insemination using frozen sperm from an abandoned bull in the ex-evacuation zone of the FNPP accident

10.2.4 Histological Analysis

The delivered calf testis was fixed in Bouin's solution. Sections were stained with haematoxylin and eosin (HE) and analysed microscopically as previously described [12].

10.2.5 Evaluation of Sperm Acrosomal Integrity

The acrosomal integrity of the delivered calf sperm was assessed by staining with fluorescein isothiocyanate-conjugated peanut agglutinin (FITC–PNA; Wako, Japan) according to the procedure described previously [13].

10.2.6 Measurement of Metabolites

Metabolome analysis of blood plasma was carried out by a facility service, Human Metabolome Technologies Inc. (Yamagata, Japan), according to the conditions described elsewhere [14–16]. A capillary electrophoresis system, Agilent 6210 Time-of-Flight LC/MS (Agilent Technologies, Waldbronn, Germany), was used for capillary electrophoresis time-of-flight mass spectrometry (CE–TOF–MS). The spectrometer scanned from m/z 50 to 1,000. Peak information included m/z, migration time (MT) and peak area. Peaks were obtained using an automatic integration software MasterHands (Keio University, Tsuruoka, Japan) [17]. Signal peaks corresponding to isotopomers, adduct ions and other product ions of known metabolites were excluded, and the remaining peaks were annotated with putative metabolites from the HMT metabolite database based on their m/z and MTs values determined by TOF–MS. The tolerance range for peak annotation was configured at ±0.5 min for MT and ±10 ppm for m/z. In addition, peak areas were normalised against those of the internal standards, and then the resultant relative area values were further normalised by the sample amount.

Hierarchical cluster analysis (HCA) and principal component analysis (PCA) were performed by our proprietary software, PeakStat and SampleStat, respectively. Detected metabolites were plotted on metabolic pathway maps using VANTED (Visualization and Analysis of Networks containing Experimental Data) software [18].

Compounds with a relative area ratio smaller than 0.66 or bigger than 1.5 were defined as having different levels between the delivered calf and the non-radiocontaminated control.

10.3   Results and Discussion

Total dose-rate in the testis of the abandoned bull attributed to radioactive Cs was 33.9 μGy/day (2.7 μGy/day for internal exposure and 31.2 μGy/day for external exposure).

Histology of the testis from the delivered calf at 13 months old revealed no remarkable changes in spermatogonia, spermatocytes, spermatids and sperm (Fig. 10.2). Sperm nuclear and acrosomal morphology was generally normal (Fig. 10.3). Abnormal sperm morphology, featuring partly absent acrosome, was observed at a rate of 7.0% (Table 10.1). These results show that, in the present study, paternal exposure to LDR radiation persistently did not affect the spermatogenesis process for calf.

CE–TOF–MS-mediated metabolic profiling of plasma resulted in 104 peaks of candidate compounds associated with exposure to chronic LDR radiation (hereafter "detected candiate" compounds). Among energy carries, AMP (ratio = 0.5) and ATP (ratio = 1.6) showed different levels between the delivered calf and the non-radiocontaminated control, whereas ADP did not (Table 10.2; Fig.10.4). Detected candidate compounds associated with the energy supply system are shown in Table 10.3 and Fig. 10.3. Regarding glycolysis, the present study showed that contents of nucleotide sugars such as G1P, G6P, DHAP and 3-PG were lower in the

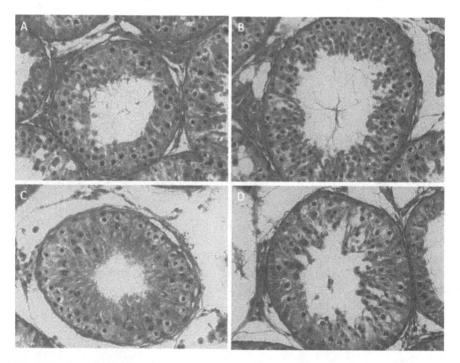

Fig. 10.2 Histological sections of the seminiferous tubules of the delivered calf testis. Scale bar, 100 μm in **a** and **d**; magnification, 200×

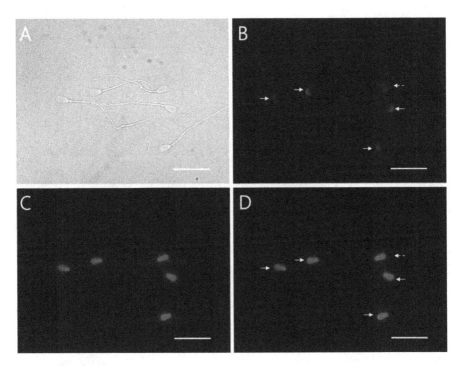

Fig. 10.3 Epididymal sperm nuclei and acrosomes of the delivered calf stained with 4′,6′-diamino-2′-phenylindole (DAPI) and fluorescein isothiocyanate-conjugated peanut agglutinin (FITC–PNA). (**a**) Phase contrast, (**b**) DAPI-stained, (**c**) FITC–PNA-stained and (**d**) DAPI+FITC–PNA-stained images. Scale bar, 50 μm; magnification, 400×

Table 10.1 Rates of sperm acrosome normal or abnormal morphology

	Sperm acrosome morphology		Total
	Normal (%)	Abnormal[a] (%)	
Number of sperm	186 (93.0)	14 (7.0)	200

[a]Abnormal sperm morphology, featuring partly absent acrosome

Table 10.2 Detected candidate energy-carrier compounds

	Relative area		Comparative analysis[a]
Energy carrier	Control calf	Delivered calf[b]	Ratio
ADP	7.0E-04	7.5E-04	1.1
AMP	4.5E-04	2.1E-04	0.5
ATP	3.4E-04	5.5E-04	1.6

[a]Control relative area served as denominator
[b]The delivered calf with paternal exposure to radiation ($n = 1$)

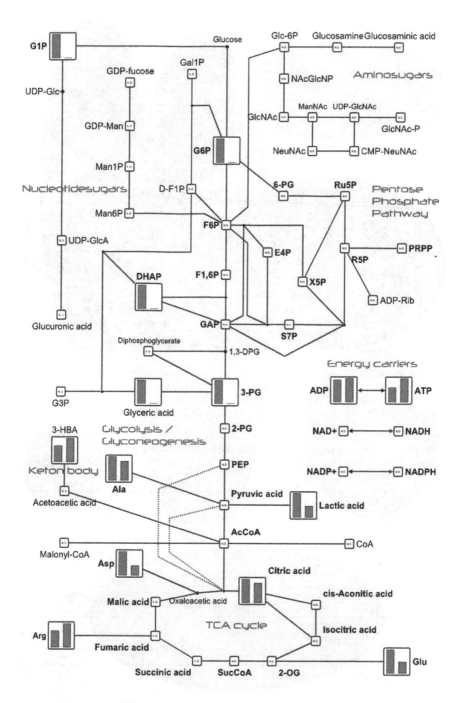

Fig. 10.4 Diagram of carbohydrate metabolism of the delivered calf (red) and the non-radiocontaminated control (blue). The control relative area was served as the denominator

Table 10.3 Detected candidate compounds involved in carbohydrate metabolism

Carbohydrate metabolite	Relative area		Comparative analysis[a]
	Control calf	Delivered calf[b]	Ratio
3-Hydroxybutyric acid	2.6E-02	3.8E-02	1.4
Citric acid	2.9E-02	2.5E-02	0.8
Lactic acid	1.1E-01	5.0E-02	0.5

[a]Control relative area served as denominator
[b]The delivered calf with paternal exposure to radiation ($n = 1$)

Table 10.4 Detected candidate amino acids

Amino acid	Relative area		Comparative analysis[a]
	Control calf	Delivered calf[b]	Ratio
Ala	5.9E-02	4.7E-02	0.8
Arg	1.9E-02	2.7E-02	1.4
Asn	5.2E-03	6.2E-03	1.2
Asp	3.2E-03	1.3E-03	0.4
Cys	N. D.		
Gln	4.8E-02	7.1E-02	1.5
Glu	2.3E-02	1.1E-02	0.5
Gly	5.0E-02	4.9E-02	1.0
His	1.6E-02	1.7E-02	1.1
Ile	5.4E-02	6.3E-02	1.2
Leu	N. D.		
Lys	1.7E-02	2.4E-02	1.4
Met	7.8E-03	6.7E-03	0.9
Phe	2.4E-02	2.3E-02	1.0
Pro	2.9E-02	2.5E-02	0.9
Ser	1.1E-02	1.4E-02	1.3
Thr	2.0E-02	1.7E-02	0.8
Trp	1.1E-02	1.3E-02	1.1
Tyr	1.5E-02	1.5E-02	1.0
Val	1.0E-01	1.1E-01	1.1

[a]Control relative area served as denominator
[b]Delivered calf associated with paternal exposure to radiation ($n = 1$)

delivered calf than in the control calf. Lactic acid, a final product of glycolysis under an aerobic conditions, featured lower-level tendency in the delivered calf than in the control calf (ratio = 0.5). 3-Hydroxybutyrate (3-HBA) is a ketone body used as an energy source, whereby it is converted to acetyl CoA and oxidised in the TCA cycle. 3-HBA showed also no difference (ratio = 1.4). Citric acid is a metabolite of the TCA cycle that showed no difference between the delivered calf and the control (ratio = 0.8). This indicates that, under anaerobic conditions, the energy use of the delivered calf has lower tendency than that of the non-radiocontamined calf.

Amino acids catabolized to simple intermediate and oxidised in the TCA cycle showed almost no difference between the delivered calf and the control (Table 10.4), with the exception of aspartic acid (Asp; ratio = 0.4), glutamine (Gln; ratio = 1.5), and glutamine acid (Glu; ratio = 0.5). Glu collects amino groups from most amino acids, including Asp, followed by conversion to α-ketoglutarate. Since Gln is converted to Glu by glutaminase, it is possible that higher levels of Gln in the delivered calf, as compared with the control calf, serve to produce Glu, which in turn activates amino acid catabolism. These may also explain the higher ATP levels detected in the delivered calf.

Arg, Asp, citrulline, and ornithine participate in the urea cycle (Table 10.5 and Fig. 10.5), a pathway that eliminates ammonia outside the body by converting it to

Table 10.5 Detected candidate compounds involved in urea metabolism

Urea metabolite	Relative area		Comparative analysis[a]
	Control calf	Delivered calf[b]	Ratio
Citrulline	1.6E-02	1.4E-02	0.9
Ornithine	1.2E-02	1.5E-02	1.3
Urea	3.8E-01	5.2E-01	1.4

[a]Control relative area served as denominator
[b]Delivered calf associated with paternal exposure to radiation ($n = 1$)

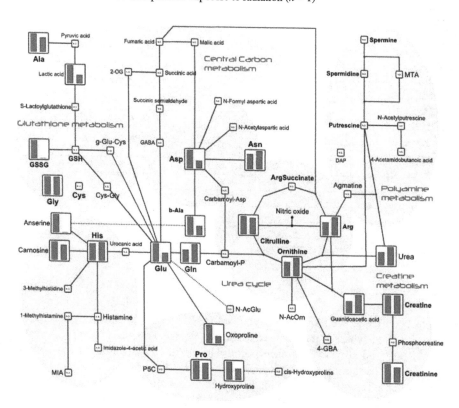

Fig. 10.5 Diagram of urea cycle of the delivered calf (red) and the non-radiocontaminated control (blue). The control relative area was served as the denominator

Table 10.6 Detected candidate compounds involved in lipid metabolism

Lipid metabolite	Relative area		Comparative analysis[a]
	Control calf	Delivered calf[b]	Ratio
Betaine	4.0E-02	4.0E-02	1.0
Carnitine	3.0E-03	5.5E-03	1.8
Choline	2.9E-02	6.9E-03	0.2
Glycerophosphocholine	1.7E-03	3.5E-04	0.2
Acetylcarnitine (ALCAR)	1.3E-03	1.5E-03	1.2
Phosphorylcholine	1.2E-03	9.1E-04	0.7
Sarcosine	9.6E-04	7.6E-04	0.8
Taurine	3.3E-03	1.8E-03	0.5

[a]Control relative area served as denominator
[b]Delivered calf associated with paternal exposure to radiation ($n = 1$)

urea. Urea levels were not different between the delivered calf and the control (ratio = 1.4), as was observed in Arg (ratio = 1.4), citrulline (ratio = 0.9) and ornithine (ratio = 1.3). Conversely, Asp showed lower levels in the delivered calf than in the control calf (ratio = 0.4). Although Asp gives a nitrogen atom to urea in the urea cycle, we assume that the difference in Asp levels had no influence on urea production in the delivered calf, since urea levels were not different between them.

Detected candidate compounds associated with lipid metabolism are shown in Table 10.6 and Fig.10.6. Carnitine is a carrier in the mitochondrial membrane responsible for transport of long-chain fatty acids into the mitochondrial matrix, where they are β-oxidised and used as the energy source in the TCA cycle. Carnitine levels were considerably higher in the delivered calf than in the control (ratio = 1.8). 5-Hydroxylysine, acetylcarnitine (ALCAR) and actinin, all associated with carnitine metabolism, were not different between them. Choline is a constituent of phosphatidylcholine and acetylcholine. It is converted to glycerophosphocholine, which is in turn converted to phosphatidylcholine. Glycerophosphocholine serves as a choline source. Choline and glycerophosphocholine were less abundant in the delivered calf than in control calf (ratio = 0.2), suggesting potential shortage in choline supply in the delivered calf as compared with the control calf. Choline is metabolised to betaine, then to sarcosine, and finally to glycine (Gly), but no difference in the levels of those metabolites, as well as that of phosphorylcholine, was observed between them. Taurine plays a role in conjugating bile acids, which are synthesised from cholesterol to form bile salts. Bile salts and acids are discharged from the body in the faeces. Taurine was less abundant in the delivered calf than in the control calf (ratio = 0.5), suggesting that bile-salt production tends to be delayed in the delivered calf.

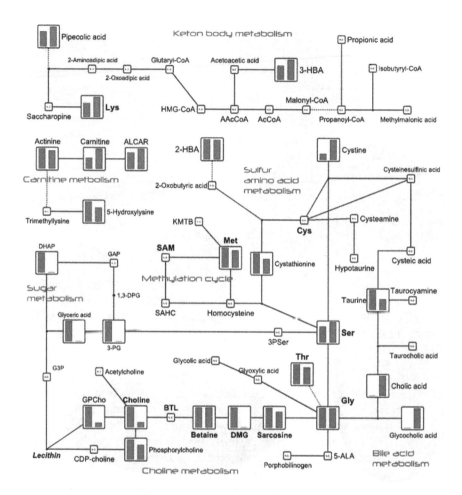

Fig. 10.6 Diagram of lipid metabolism of the delivered calf (red) and non-radiocontaminated control (blue). The control relative area was served as the denominator

10.4 Conclusions

In conclusion, there were no transgenerational effects on both of the spermatogenesis and on the blood plasma metabolome in calf obtained by AI using sperm from a LDR-radiated bull in the FNPP ex-evacuation zone. The area of the evacuation zone has been drastically reduced, and no more cattle have been euthanized since February 2014. It is therefore, difficult to investigate more animals with different annual doses. Here, we showed results of only one 13-month-old calf obtained by AI using sperm from a bull exposed to chronic LDR radiation for 2 years in the ex-evacuation zone of the FNPP accident (Fig. 10.1). This study is not a controlled experiment; however, we believe that our results give a framework not only for estimating general effects of radiation in cattle but also contributing to the reconstruction of the Fukushima livestock industry and facilitating the improvement of food safety measures.

Acknowledgements We express our gratitude to the Iwaki Livestock Hygiene Service Centre in Fukushima Prefecture, Japan. We thank DVM Yuji Kobayashi and livestock farmers in the 20-km FNPP ex-evacuation zone. We thank DVM Taro Sato at Taro Veterinary Farm, Niigata. This work was partly supported by a grant from the Japan Society for the Promotion of Science (H.Y 15K11952 and M.F. 26253022) and The Programme for Promotion of Basic and Applied Researches for Innovations in Bio-oriented Industry.

Conflict of Interest Statement The authors declare no competing financial interests.

References

1. Kinoshita N, Sueki K, Kitagawa J et al (2013) Assessment of individual radionuclide distributions from the Fukushima nuclear accident covering central-east Japan. Proc Natl Acad Sci U S A 108:19526–19529
2. Takahashi S, Inoue K, Urushihara Y et al (2015) A comprehensive dose evaluation project concerning animals affected by the Fukushima Daiichi nuclear power plant accident: its setup and progress. J Radiat Res 56(S1):i36–i41
3. Urushihara Y, Kawasumi K, Endo S et al (2016) Analysis of plasma protein concentrations and enzyme activities in cattle within the ex-evacuation zone of the Fukushima Daiichi nuclear plant accident. PLoS One 11:e0155069
4. Koarai K, Kino Y, Takahashi A et al (2016) ^{90}Sr in teeth of cattle abandoned in evacuation zone: record of pollution from the Fukushima-Daiichi nuclear power plant accident. Sci Rep 6:24077
5. Takino S, Yamashiro H, Sugano Y et al (2016) Analysis of the effect of chronic and low-dose radiation exposure on spermatogenic cells of male large Japanese field mice (Apodemus speciosus) after the Fukushima Daiichi nuclear power plant accident. Radiat Res 187:161–168
6. Fukuda T, Kino Y, Abe Y et al (2013) Distribution of artificial radionuclides in abandoned cattle in the evacuation zone of the Fukushima Daiichi nuclear power plant. PLoS One 8:e54312
7. Fukuda T, Kino Y, Abe Y et al (2015) Cesium radioactivity in peripheral blood is linearly correlated to that in skeletal muscle: analyses of cattle within the evacuation zone of the Fukushima Daiichi nuclear power plant. Anim Sci J 86:120–124
8. Fukuda T, Hiji M, Kino Y et al (2016) Software development for estimating the cesium radioactivity in skeletal muscles of cattle from blood samples. Anim Sci J 87:842–847
9. Yamashiro H, Abe Y, Fukuda T et al (2013) Effects of radioactive caesium on bull testes after the Fukushima nuclear plant accident. Sci Rep 3:2850
10. Yamashiro H, Abe Y, Hayashi G et al (2015) Electron probe X-ray microanalysis of boar and inobuta testes after the Fukushima accident. J Radiat Res 56(S1):i42–i47
11. Yamashiro H, Abe Y, Kuwahara Y et al (2014) Cryopreservation of cattle, pig, inobuta sperm and oocyte after the Fukushima nuclear plant accident. In: Recent advances in cryopreservation. IntechOpen, pp 73–81
12. Akiyama M, Takino S, Sugano Y et al (2015) Effect of seasonal changes on testicular morphology and the expression of circadian clock genes in Japanese wood mice (Apodemus speciosus). J Biol Regul Homeostat Agent 29:589–600
13. Kaneko K, Uematsu E, Takahashi Y et al (2013) Semen collection and polymerase chain reaction-based sex-determination of black-headed and straw-necked ibis. Reprod Domest Anim 48:1001–1005

14. Soga T, Heiger HD (2000) Amino acid analysis by capillary electrophoresis electrospray ionization mass spectrometry. Anal Chem 72:1236–1241
15. Soga T, Ueno Y, Naraoka H et al (2002) Simultaneous determination of anionic intermediates for Bacillus subtilis metabolic pathways by capillary electrophoresis electrospray ionization mass spectrometry. Anal Chem 74:2233–2239
16. Soga T, Ohashi Y, Ueno Y et al (2003) Quantitative metabolome analysis using capillary electrophoresis mass spectrometry. J Proteome Res 2:488–494
17. Sugimoto M, Wong DT, Hirayama A et al (2010) Capillary electrophoresis mass spectrometry-based saliva metabolomics identified oral, breast and pancreatic cancer-specific profiles. Metabolomics 6:78–95
18. Junker BH, Klukas C, Schreiber F (2006) A system for advanced data analysis and visualization in the context of biological networks. BMC Bioinformatics 7:109

Pigs Immune System and the Impact of FNPP Accident

Motoko Morimoto, Ayaka Kato, Jin Kobayashi, Kei Okuda,
Yoshikazu Kuwahara, Yasushi Kino, Yasuyuki Abe, Tsutomu Sekine,
Tomokazu Fukuda, Emiko Isogai, and Manabu Fukumoto

Abstract It is extremely important to understand the mechanisms underlying the effect of radiation on animals including humans. Radiation potentially induces alterations of the immune system, which may cause serious diseases. The Fukushima Daiichi Nuclear Power Plant (FNPP) accident occurred in 2011, in which radioactive contaminants were released over a wide area. After farmers evacuated from the

M. Morimoto (✉) · A. Kato · J. Kobayashi
School of Food, Agricultural, and Environmental Sciences, Miyagi University,
Sendai, Japan
e-mail: morimoto@myu.ac.jp

K. Okuda
Institute of Environmental Radioactivity, Fukushima University, Fukushima, Japan

Y. Kuwahara
Department of Pathology, Institute of Development, Aging and Cancer, Tohoku University,
Sendai, Japan

Y. Kino
Department of Chemistry, Tohoku University, Sendai, Japan

Y. Abe
Faculty of Life and Environmental Sciences, Prefectural University of Hiroshima,
Hiroshima, Japan

T. Sekine
Institute for Excellence in Higher Education, Tohoku University, Sendai, Japan

T. Fukuda
Graduate School of Science and Engineering, Iwate University, Morioka, Japan

E. Isogai
Graduate School of Agricultural Sciences, Tohoku University, Sendai, Japan

M. Fukumoto
Department of Pathology, Institute of Development, Aging and Cancer, Tohoku University,
Sendai, Japan

School of Medicine, Tokyo Medical University, Tokyo, Japan

ex-evacuation zone set within 20-km radius from FNPP, pigs were left behind and unleashed in the zone (abandoned pigs). Their small intestine was considered to have been affected by the dietary intake of radioactive contaminants. The present study thus aimed to clarify the effect of radiation by investigating whether there is any alteration in the expression of genes encoding immune-related molecules and in morphology of the intestine of abandoned pigs. Microarray analysis revealed changes of the expression of various genes involved in inflammation and oxidative stress including IFN-γ, which is an important inflammatory cytokine, was highly elevated in abandoned pigs, and its expression level was correlated with radioactive cesium concentration in skeletal muscle. On the other hand, there were no morphological changes in the intestine. At the sampling stage of this study, only 1 year passed since the FNPP accident occurred, it would be premature to conclude that the identified alteration of gene expression was caused only by artificial radionuclides attributed to the FNPP accident. We need to continue monitoring the effect of long-term radiation exposure in living organisms in the ex-evacuation zone.

Keywords Small intestine · Immune responses · Gene expression · Plasma biochemistry · Swine

11.1   Introduction

The immune system is essential not only for host defense against pathogens but also for regulation of physiological functions. The immune cells and nonimmune cells cooperate to maintain homeostasis in the body. Their interaction is regulated by a range of biologically active substances called cytokines. A large number of cytokines have been found to play key roles in regulating immune responses and inflammation. The immune system consists of innate immunity, which mediates the initial reaction by neutrophils, macrophages and innate lymphocytes (ILCs), and adaptive immunity, which develops more slowly but produces a specific reaction against antigens by lymphocytes [1]. The lymphocytes, T and B lymphocytes, are the most important immune cells for the development of appropriate and effective reactions in various situations. Inappropriate immune responses, including cytokine imbalances, contribute to the development of persistent diseases [2]. Specifically, insufficient responses result in the establishment of infection, while excessive activation of the immune system may cause chronic inflammation, which is central to the pathogenesis of various diseases [3].

A large number of studies on the effect of radiation on the immune system have been performed in animal models. As mentioned above, disorders of the immune system are closely associated with several diseases, including cancer, neurodegeneration and inflammatory diseases. Therefore, it is important to understand the effect of radiation on the immune system. UNSCEAR 2006 Report summarized the details of the effect of radiation on the immune system as described in the literature [4]. Although no definitive conclusions have yet been drawn, it was suggested that radiation induces a shift toward an inflammatory profile that could increase the risk of diseases. Radiation-induced oxidative stress affects biomolecules in immune cells, which probably leads to abnormal T-cell differentiation and functions, promoting or controlling acute and chronic diseases. In other words, the interaction of radiation with the water molecules in living cells induces the production of free radicals. In terms of the consequences of this, it is well known that reactive oxygen species (ROS) activate nuclear factor kappa B (NF-κB) which regulates the expression of a large number of target genes involved in immune and inflammatory responses [5]. Chronic inflammation not only promotes carcinogenesis, malignant transformation, tumor growth, invasion and metastatic spread, but also regulates the immune system, which limits tumor growth [6]. At an inflammatory site, several immune cells produce inflammatory cytokines. The function of cytokines is to induce the proliferation, survival and growth of cells and modify their differentiation status. Thus, the cytokine profile may be a useful marker of the alteration of immune responses. Hayashi et al. found that TNF-α, IFN-β and IL-6 increase in atomic bomb survivors (Hibakusha) [7]. Moreover, Matsuoka et al. found increased IL-6 after radiotherapy, which is associated with radioresistance [8]. Furthermore, Lugade et al. learned that IFN-γ is upregulated following radiation in a melanoma model [9]. Recently, it has been suggested that this elevation of inflammatory cytokines makes an important contribution to radioresistance or radiosensitivity, as well as inducing or suppressing radiation-mediated immune responses [10]. However, most of the obtained data were collected from culture cells, and sufficient data from the field and in vivo are not available.

After the Great East Japan Earthquake on March 11, 2011, a huge amount of radioactive cesium (Cs) was released following the Fukushima Daiichi Nuclear Power Plant (FNPP) accident. After farmers evacuated from the ex-evacuation zone set within a 20-km radius from FNPP, pigs were left behind and unleashed in the zone (abandoned pigs). We have constructed "the Group for Comprehensive Dose Evaluation in Animals from the Area Affected by the FNPP Accident" and have been continuing to monitor the effect of long-term radiation exposure among organisms living in the ex-evacuation zone. The immune system and physiological functions of pigs are very similar to those of humans [11–13]. Therefore, knowledge about the immunological responses of abandoned pigs to radioactive contamination can be useful to understand the effect of radiation on humans. Here, we report our data of abandoned pigs and discuss the consequence of the FNPP accident.

11.2 Methods

11.2.1 Animals

We collected peripheral blood from 53 euthanized pigs and obtained the intestinal and the skeletal muscle samples from 13 of them between January 18 and February 16, 2012, 5 km southwest of FNPP. The animals were sacrificed in accordance with the Regulations for Animal Experiments and Related Activities, Tohoku University, by veterinary doctors belonging to the Livestock Hygiene Service Center of Fukushima Prefecture [14]. Estimates of the amounts of [134]Cs and ^{137}Cs deposited on the ground have been reported [15]. Control intestine samples were obtained from three healthy pigs present in an uncontaminated pigsty in Miyagi Prefecture in 2012. Each experimental protocol was approved by the Institutional Ethics Commissions for Animal Research at Tohoku University and Miyagi University.

11.2.2 Measurement of Radioactivity

Radioactivity in the muscle samples was determined by γ-ray spectrometry using high-purity germanium (HPGe) detectors (Ortec Co., Oak Ridge, TN, USA), as described in our previous report [14]. Gamma-rays from [134]Cs and ^{137}Cs were observed and radioactivity ratios of [134]Cs to [137]Cs (decay corrected to March 15, 2011) were 0.9–1.0, which corresponded to those of other samples polluted by the FNPP accident.

11.2.3 Histological Analysis

Small pieces of the small intestine were slit longitudinally, laid flat with the mucosal surface facing down and rolled around a wooden stick (Swiss roll). Paraffin blocks were prepared for pathomorphological examination using hematoxylin and eosin (HE) staining.

11.2.4 Gene Expression Analysis

Total RNA was extracted from the small intestine collected in 2012 using TRIzol reagent (Life Technologies, Inc., Frederic, MD, USA), in accordance with the manufacturer's instruction. RNA concentration was measured on a NanoDrop spectrophotometer (Thermo Scientific, Wilmington, DE, USA) and cDNA was synthesized with random primers and SuperScript II (Life Technologies, Inc.). cDNA samples were analyzed using a microarray (three pigs of control vs three abandoned pigs in the ex-evacuation zone) (V1: 4 × 44K; Agilent Technologies, Palo Alto, CA, USA), and IFN-γ gene expression was assessed by real-time PCR. Primer sequences were designed using Primer-BLAST with sequences obtained from GenBank [16].

Real-time PCR was performed using Brilliant SYBR Green QPCR Master Mix III (Stratagene, La Jolla, CA, USA) with an MX3000P system (Stratagene, La Jolla, CA, USA). Amplification conditions were as follows: 95 °C for 3 min, 40–50 cycles at 95 °C for 5 s and 60 °C for 20 s. Fluorescence signals measured during the amplification were analyzed. Ribosomal RNA primers were used as an internal control and all data were normalized to constitutive rRNA values. Quantitative differences between the groups were calculated in accordance with the manufacturer's instructions as expression ratio (Applied Biosystems, Foster City, CA, USA).

11.2.5 Biochemical Testing

Heparinized peripheral blood was collected from the jugular vein and immediately centrifuged to separate plasma and blood cells. Plasma samples were preserved at −80 °C until use. Biochemical testing was performed with the Fuji Dri-Chem system (NX500; Fuji Film, Tokyo, Japan) to measure glutamic oxaloacetic transaminase (GOT), blood urea nitrogen (BUN) and total cholesterol (TCHO).

11.2.6 Statistical Analysis

All data are presented as mean ± standard error (SE) for each treatment group. Differences in mRNA expression among the groups were determined using the t-test (Prism; GraphPad Software Inc., La Jolla, CA, USA). Differences were considered statistically significant at a p-value of <0.05.

11.3 Results

11.3.1 Living Conditions of Pigs in the Ex-Evacuation Zone

When our team started to investigate the effect of radiation on animals around and in the ex-evacuation zone, most of the abandoned pigs were unleashed. They were eating, sleeping and walking around outside the barn without restrictions, although they may not have had sufficient access to concentrated feed. However, the pigs looked healthy and clean, and no emaciation was observed (Fig. 11.1a), although some of them were infected by *Trichuris suis* (Fig. 11.1b), a typical whipworm in swine. Nonetheless, *T. suis* does not cause serious disease in adult pigs unless the host has an extremely high worm load.

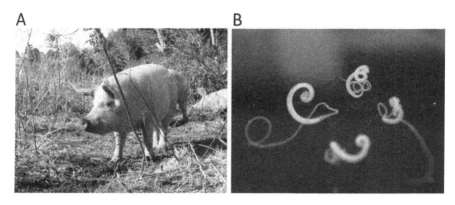

Fig. 11.1 (**a**) An abandoned unleashed pig in the ex-evacuation zone in 2012. (**b**) *Trichuris suis* collected from the pig cecum

11.3.2   Microarray Analysis

To identify the genes that were up- or down-regulated by radiation in the ex-evacuation zone, RNA was extracted for cDNA microarray analysis from the small intestine, which is highly sensitive to radiation [17]. The number of differentially expressed genes with a ≥2-fold change was 5,135. Various immune-related genes were up-regulated in abandoned pigs caught in 2012 [16]. Table 11.1 shows upregulated genes that are involved in NF-κB activation, oxidative stress and inflammatory immune responses. We found that many genes involved in responses to oxidative stress were up-regulated.

11.3.3   Real-Time PCR Analysis

The alterations in gene expression could be considered as the evidence of changes in physiological function after exposure to radiation. We have already reported that some of the selected genes did not show any changes; however, some showed significant differences. As previously reported [16], the expression levels of the AIFM1, IFN-γ and TLR3 genes were significantly higher in abandoned pigs than in control pigs. However, the expression levels of cyclin G1, GADD45A, XRCC1, Smad7, XAB2, XPC, OGG1 and EPHX2 were similar in both groups. Here, we reprint the results of expression of the IFN-γ gene obtained in our previous study in Fig. 11.2a. The expression ratio and the elevation of IFN-γ showed large interindividual differences. Therefore, we next focused on the relationship between gene expression and ^{137}Cs concentration because if radiation causes inflammation through oxidative stress, the elevation of IFN-γ expression might be related to the level of radiation exposure of each pig. Figure 11.2b is a modified version of a figure from our previous report [16]. IFN-γ gene expression was associated with ^{137}Cs concentration in the muscle ($y = 0.3687 \times 10^{-6} \pm 0.1342 \times 10^{-6}\, x + 0.05727 \pm 0.002299\, [R^2 = 0.3502]$).

Table 11.1 The number of significantly differentially expressed genes (>2-fold) linked to immune responses by microarray data analysis

Category	The number of ≥2-fold change genes in category	p-Value
Cytokine production involved in inflammatory response	2	0.034
Negative regulation of NF-kappaB transcription factor activity	8	0.004
Negative regulation of cytokine production involved in inflammatory response	2	0.034
Regulation of cytokine production involved in inflammatory response	2	0.034
Regulation of intrinsic apoptotic signaling pathway in response to oxidative stress	3	0.021
Positive regulation of intrinsic apoptotic signaling pathway in response to oxidative stress	3	0.011
Positive regulation of response to oxidative stress	3	0.011
Regulation of response to oxidative stress	3	0.021
Response to oxidative stress	28	0.007
Cellular response to oxidative stress	17	0.014
Positive regulation of cellular response to oxidative stress	3	0.011
Regulation of cellular response to oxidative stress	3	0.021

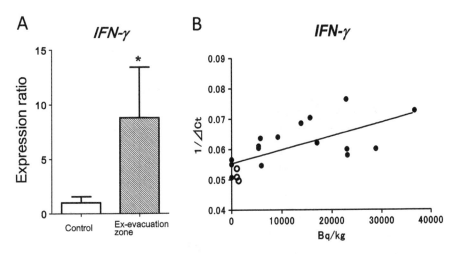

Fig. 11.2 (a) Real-time PCR analysis of IFN-γ gene expression in the small intestine of control pigs ($n = 3$) and pigs from the ex-evacuation zone in 2012 ($n = 13$). All data are expressed in relative units compared with control pigs. *$P < 0.05$. Data are presented as mean ± S.E. B: [137]Cs concentration in skeletal muscle and expression of each gene (black circle, pigs in the ex-evacuation zone; white circle, control). A positive correlation was observed between muscle radioactivity and expression of IFN-γ ($R^2 = 0.3502$). (a) Is reprinted and (b) is modified from our previous report [16]

11.3.4 Plasma Biochemistry

Plasma biochemistry refers to the chemical analysis of blood plasma which reflects the function of crucial organs. Most of the abandoned pigs examined (49/53) showed within the normal levels of GOT and the median value was also in the normal range (Fig.11.3a). Four out of 53 abandoned pigs showed high levels of GOT, which could be attributed to parasitic infection as a possible explanation. Adult *Ascaris suum* is a common roundworm in pigs worldwide. The migration of the larvae of this species through the liver causes parasite-induced liver inflammation. However, we did not find *Ascaris suum* in abandoned pigs.

Blood urea nitrogen (BUN) reflects the kidney function. The BUN value of half of the abandoned pigs was outside the normal range (Fig.11.3b). The detailed reason is not clear, but elevated BUN might be due to dehydration resulting from not drinking enough, and lower BUN might be due to the poor nutritional condition, especially low-protein diet.

Total cholesterol (TCHO) is an indicator of nutritional condition. About half of the abandoned pigs examined showed levels of TCHO below the normal range (Fig.11.3c), which could be attributed to a lack of concentrated feed.

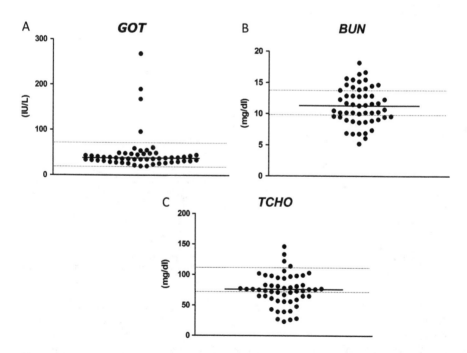

Fig. 11.3 Biochemical analysis of peripheral blood from abandoned pigs. (**a**) *GOT*: glutamic oxaloacetic transaminase; (**b**) *BUN*: blood urea nitrogen; (c) *TCHO*: total cholesterol (*n* = 53). Normal ranges are indicated with dashed lines

11.3.5 Pathological and Morphological Changes in the Small Intestine

As we previously reported that there were no remarkable changes including fibrosis in the intestinal tissues of abandoned pigs [16]. Despite the highly elevated IFN-γ gene expression, there were no pathological or morphological changes in both the abandoned and the non-affected control groups (Fig.11.4).

11.4 Discussion

The Great East Japan Earthquake and tsunami struck the Tohoku (east-north) area of Japan in 2011. The Japanese authorities eventually recategorized the situation at FNPP as a level 7 incident on the International Nuclear Event Scale. A maximum of 164,865 people were evacuated from the ex-evacuation zone, within a 20-km radius from FNPP.

Before the earthquake, livestock farming was thriving in this area. More than 184,000 pigs and 74,000 beef cattle were raised in the whole of Fukushima Prefecture in February 2011 [18]. However, the livestock were abandoned in the

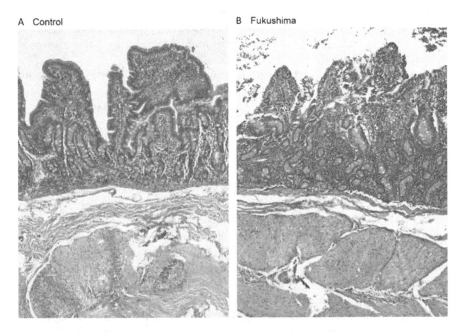

Fig. 11.4 Representative images of hematoxylin and eosin staining. (**a**) Control; (**b**) an abandoned pig. All images were acquired from 4-μm-thick sections at 40×. The results suggest no morphologically remarkable changes in the small intestine of abandoned pigs after the accident

ex-evacuation zone and most of them have already been euthanized. Our goal here was to accurately analyze the effect of radiation on the animals, to archive all of the data and to present them in a useful format for future generations. This is because such rare data from a contaminated area are very important for understanding biological reactions to radiation in vivo.

As mentioned above, the immune system is extremely important for maintaining homeostasis and tissue repair during radiation stress. However, many conflicting results have been obtained from researches on the interaction between ionizing radiation and the immune system. As Tang et al. stated in their review, various animal models have been investigated in complex conditions to assess the bioeffects of low-dose or low-dose-rate (LD/LDR) radiation [19]. The mechanisms behind LD/LDR radiation-induced alterations differ depending on the animal species, strain, age, sex and organ. In an investigation on immunological effects, C57BL/6 mice exposed to 0.01–3.0 Gy of X-rays demonstrated a higher level of TNF-α at 24 h after irradiation, while the elevation of TGF-β expression was identified in CBA mice [20]. TGF-β expression was also highly increased at 3 days postirradiation in Balb/c mice irradiated with 0.1–5 Gy of γ-irradiation [21]. TGF-β has antiproliferative and cell cycle regulatory activities and is also well known to play an important role in genomic stability. TGF-β is also known as a unique cytokine that acts in inducing regulatory T cells (Tregs), which inhibit excessive immune responses and in tissue repair to induce fibrosis. TNF-α is involved in inflammatory responses linked to NF-kB activation. TNF-α and its signaling molecules are key regulators of ROS/oxidative-induced cell death [22]. The alteration of TNF-α and TGF-β levels has been observed after exposure to LD radiation in different model systems. Therefore, the monitoring of inflammatory molecules in abandoned animals would be useful to assess the effect of the FNPP accident.

As shown in Fig. 11.1a, the abandoned pigs were living freely and thus could have consumed contaminated insects, soil, water and plants. Therefore, we considered that the effect of radiation might have appeared in the gastrointestinal tract. It is well known that the gastrointestinal tract is highly sensitive to radiation. The intestine has simple columnar epithelium of which cells are tied together with tight junctions, making it a robust barrier to protect the mucosa. The expression of adhesion factors, such as zo-1, zo-2 and afadin, which are scaffold proteins, has also been reported to be altered by radiation [23]. Once disruption of the mucosal barrier has occurred, the gut immune system is activated. The gut mucosa is the largest and the most dynamic immunological environment in the body. Veeraraghave et al. observed NF-κB activation in the gut after 10–50 cGy of γ-radiation, which in turn causes the up-regulation of several genes [24]. In this study, we also found the up-regulation of genes involved in NF-κB activation and oxidative stress in the gut of abandoned pigs, through microarray analysis (Table 11.1). Yu et al. demonstrated that LDR radiation promotes dendritic cell migration and IL-12 secretion through the NF-κB pathway [25]. IL-12 is one of the most important cytokines for the initiation of inflammatory responses. As we have already described, quantitative gene expression analysis revealed that IFN-γ gene expression was up-regulated in abandoned pigs. In addition, there was a significant correlation between ^{137}Cs concentration, that is, dose-rate in skeletal muscle and IFN-γ gene expression. However, plasma biochemistry suggested that most of the abandoned pigs examined did not have any problems with their health. According to the data of GOT, some of the abandoned pigs might have been infected with *A. suum* in the past. The larvae of this species migrate from the cecum to the liver, resulting in parasitic hepatitis [26]. In addition, as shown in Fig. 11.1b, some abandoned pigs were infected with *T. suis*, which does not normally cause severe symptoms in the host. In general, nematode infection usually induces Th2 cytokines in the host [27–29], but we did not detect any elevation of IL-4 and IL-10 in abandoned pigs (data not shown). Therefore, we considered that the increase of IFN-γ gene expression in abandoned pigs was not due to parasitic infection. As mentioned above, IFN-γ is one of the inflammatory cytokines acting in host defense against viral and bacterial infections. Moreover, IFN-γ has pleiotropic effects, including the inhibition of cell proliferation, angiogenesis and apoptosis; in other words, it exerts antitumor effects [30–32]. Zaidi and Merlino stated that the effect of IFN-γ is dependent on the cellular, microenvironmental and/or molecular context [33]. As shown in Fig. 11.4, no pathological changes in the intestinal tissue were observed and the animals did not show any abnormal health condition. Therefore, the elevation of IFN-γ in abandoned pigs is thought to have been due to host defense against radiation, which could work to maintain the physiological functions in a harsh environment. A previous study demonstrated that the guanine nucleotide-binding protein 1 (GBP1) is one of the genes most strongly induced by interferons. Its overexpression is necessary for the clinically relevant radioresistant phenotype (resistant against persistent exposure to 2 Gy/day of fractionated X-rays) in human cancer cells through inhibiting apoptosis. Thus, IFN-γ gene overexpression in this study suggests that adaptive response against LD/LDR radiation is induced in small intestine of pigs in the ex-evacuation zone [34]. However, at present, we cannot draw any definitive conclusions on whether the elevation of IFN-γ represents a form of protection or damage in the host. More analyses are required to elucidate the effect of long-term LDR radiation exposure in living organisms. We thus need to continue monitoring wild animals in the ex-evacuation zone.

References

1. Abbas AK, Lichtman AH, Pillai S (2017) Cellular and molecular immunology. Elsevier, Philadelphia
2. Muller B (2002) Cytokine imbalance in non-immunological chronic disease. Cytokine 18(6):334–339
3. Chizzolini C, Dayer JM, Miossec P (2009) Cytokines in chronic rheumatic diseases: is everything lack of homeostatic balance? Arthritis Res Ther 11(5):246
4. United Nations Scientific Committee on the Effects of Atomic Radiation: UNSCEAR 2006 Report Volume II (2006) Effects of ionizing radiation. In: Effects of ionizing radiation on the immune system

5. Ghosh S, Hayden MS (2008) New regulators of NF-kappaB in inflammation. Nat Rev Immunol 8(11):837–848
6. Multhoff G, Molls M, Radons J (2011) Chronic inflammation in cancer development. Front Immunol 2:98
7. Hayashi T, Kusunoki Y, Hakoda M et al (2003) Radiation dose-dependent increases in inflammatory response markers in A-bomb survivors. Int J Radiat Biol 79(2):129–136
8. Matsuoka Y, Nakayama H, Yoshida R et al (2016) IL-6 controls resistance to radiation by suppressing oxidative stress via the Nrf2-antioxidant pathway in oral squamous cell carcinoma. Br J Cancer 115(10):1234–1244
9. Lugade AA, Sorensen EW, Gerber SA et al (2008) Radiation-induced IFN-gamma production within the tumor microenvironment influences antitumor immunity. J Immunol 180(5):3132–3139
10. Gerber SA, Sedlacek AL, Cron KR et al (2013) IFN-gamma mediates the antitumor effects of radiation therapy in a murine colon tumor. Am J Pathol 182(6):2345–2354
11. Bode G, Clausing P, Gervais F et al (2010) The utility of the minipig as an animal model in regulatory toxicology. J Pharmacol Toxicol Methods 62(3):196–220
12. Dawson HD, Beshah E, Nishi S et al (2005) Localized multigene expression patterns support an evolving Th1/Th2-like paradigm in response to infections with toxoplasma gondii and Ascaris suum. Infect Immun 73(2):1116–1128
13. Swindle MM, Makin A, Herron AJ et al (2012) Swine as models in biomedical research and toxicology testing. Vet Pathol 49(2):344–356
14. Fukuda T, Kino Y, Abe Y et al (2013) Distribution of artificial radionuclides in abandoned cattle in the evacuation zone of the Fukushima Daiichi nuclear power plant. PLoS One 8(1):e54312
15. Yamashiro H, Abe Y, Hayashi G et al (2015) Electron probe X-ray microanalysis of boar and inobuta testes after the Fukushima accident. J Radiat Res 56(Suppl 1):i42–i47
16. Morimoto M, Kato A, Kobayashi J et al (2017) Gene expression analyses of the small intestine of pigs in the ex-evacuation zone of the Fukushima Daiichi Nuclear Power Plant. BMC Vet Res 13(1):337
17. Somosy Z, Horvath G, Telbisz A et al (2002) Morphological aspects of ionizing radiation response of small intestine. Micron 33(2):167–178
18. http://www.maff.go.jp/e/index.html
19. Tang FR, Loke WK, Khoo BC (2017) Low-dose or low-dose-rate ionizing radiation-induced bioeffects in animal models. J Radiat Res 58(2):165–182
20. Irons SL, Serra V, Bowler D et al (2012) The effect of genetic background and dose on nontargeted effects of radiation. Int J Radiat Biol 88(10):735–742
21. Ehrhart EJ, Segarini P, Tsang ML et al (1997) Latent transforming growth factor beta1 activation in situ: quantitative and functional evidence after low-dose gamma-irradiation. FASEB J 11(12):991–1002
22. Shen HM, Lin Y, Choksi S et al (2004) Essential roles of receptor-interacting protein and TRAF2 in oxidative stress-induced cell death. Mol Cell Biol 24(13):5914–5922
23. Somosy Z, Bognar G, Horvath G et al (2003) Role of nitric oxide, cAMP and cGMP in the radiation induced changes of tight junctions in Madin-Darby canine kidney cells. Cell Mol Biol (Noisy-le-Grand) 49(1):59–63
24. Veeraraghavan J, Natarajan M, Herman TS et al (2011) Low-dose gamma-radiation-induced oxidative stress response in mouse brain and gut: regulation by NFkappaB-MnSOD cross-signaling. Mutat Res 718(1–2):44–55
25. Yu N, Wang S, Song X et al (2018) Low-dose radiation promotes dendritic cell migration and IL-12 production via the ATM/NF-kappaB pathway. Radiat Res 189:409
26. Douvres FW, Tromba FG, Malakatis GM (1969) Morphogenesis and migration of Ascaris suum larvae developing to fourth stage in swine. J Parasitol 55(4):689–712
27. Gause WC, Ekkens M, Nguyen D et al (1999) The development of CD4+ T effector cells during the type 2 immune response. Immunol Res 20(1):55–65

28. Morimoto M, Saito C, Muto C et al (2015) Impairment of host resistance to helminthes with age in murine small intestine. Parasite Immunol 37(4):171–179

29. Urban JF Jr, Madden KB, Svetic A et al (1992) The importance of Th2 cytokines in protective immunity to nematodes. Immunol Rev 127:205–220

30. Beatty G, Paterson Y (2001) IFN-gamma-dependent inhibition of tumor angiogenesis by tumor-infiltrating CD4+ T cells requires tumor responsiveness to IFN-gamma. J Immunol 166(4):2276–2282

31. Chawla-Sarkar M, Lindner DJ, Liu YF et al (2003) Apoptosis and interferons: role of interferon-stimulated genes as mediators of apoptosis. Apoptosis 8(3):237–249

32. Coughlin CM, Salhany KE, Gee MS et al (1998) Tumor cell responses to IFNgamma affect tumorigenicity and response to IL-12 therapy and antiangiogenesis. Immunity 9(1):25–34

33. Zaidi MR, Merlino G (2011) The two faces of interferon-gamma in cancer. Clin Cancer Res 17(19):6118–6124

34. Fukumoto M, Amanuma T, Kuwahara Y et al (2014) Guanine nucleotide-binding protein 1 is one of the key molecules contributing to cancer cell radioresistance. Cancer Sci 105(10):1351–1359

12

A Study of Farm Animals: DNA damage due to Radiation

Asako J. Nakamura

Abstract Since the Fukushima Daiichi Nuclear Power Plant (FNPP) accident is a radiation accident which occurred as the aftermath of a devastating natural disaster, the Great East Japan Earthquake, people had to face radiation risks in the chaotic state. In the situation where general citizens, scientists and politicians were all confused, the importance of accurately assessing biological risks of radiation emerged. To understand the biological effects of radiation exposure by the accident, we measured the DNA damage level using the DNA double-strand break (DSB) marker phosphorylated histone H2AX in peripheral blood lymphocytes from farm animals left behind within a 20-km radius from FNPP (the ex-evacuation zone). As a result, statistically higher levels of DNA DSBs were detected from cattle in the ex-evacuation zone compared to non-affected control; however, it was not able to accurately evaluate the radiation dose from this accident with phosphorylated H2AX. This is thought to be caused by the fact that various changes of metabolism with the lapse of time and the living environment of individual organisms occurred. It may, therefore, be difficult to evaluate the exposure dose of chronic low-dose-rate (LDR) radiation by a single biomarker. However, in the inevitable modern society of radiation exposure and fear of nuclear accidents, our results showed that trying a certain dosimetric biomarker for the assessment of biological impacts of long-term LDR radiation exposure is effective and crucial for the protection from radiation.

Keywords Biodosimetry · DNA double-strand break · Phosphorylated histone H2AX · Biological effect

A. J. Nakamura (✉)
Department of Biological Science, College of Science, Ibaraki University, Mito, Ibaraki, Japan
e-mail: asako.nakamura.wasabi@vc.ibaraki.ac.jp

12.1 Introduction

In 1998, it was first reported that histone H2AX, one of the core histones, is rapidly phosphorylated (phosphorylated histone H2AX is called γ-H2AX) at the time of induction of DNA double-strand breaks (DSBs) and plays an important role for recruitment of DNA damage repair proteins to the damage site [1–5]. As phosphorylation of H2AX occurs specifically at the site of DNA DSBs, it is possible to visualize the DNA DSB site as the focus of γ-H2AX by immunofluorescent staining method using an antibody specific to γ-H2AX [2, 5–7]. In fact, it was shown that the γ-H2AX focus number and the number of DNA DSBs are the same [8]. Though DNA DSBs had been detected by experimental methods such as the comet assay and pulsed-field gel electrophoresis previously, γ-H2AX is sensitive enough to detect radiation exposure equivalent to several mGy, and the operation is simple; therefore, it is widely used as a marker of DNA DSB [6, 9, 10]. Recently, the γ-H2AX assay is used as a method of monitoring in vivo DNA DSB level for clinical trials of novel anticancer drugs and for assessment of exposure dose of cancer patients who received radiation therapy [11–17]. Present when this manuscript is being written (June 2018), more than 40 clinical trials using the γ-H2AX assay seem to be performed in the United States (https://clinicaltrials.gov/). Since the September 11 attacks occurred in the United States in 2001, the risk of unexpected radiation exposure increased, and the possible application of the γ-H2AX assay as a biological biomarker for correctly evaluating exposure dose has also been studied [18–20]. However, the assessment of radiation dose by the γ-H2AX assay still has problems in case of a large-scale disaster like the FNPP accident because current method does not allow us to monitor the DNA DSB level by γ-H2AX assay on site, such as at a radiation accident site. On March 11, 2011, the FNPP accident occurred, and human beings faced an unexpected radiation accident. To understand biological effects of radiation exposure due to the accident, we measured the DNA DSB level based on the γ-H2AX assay in biospecimens obtained from farm animals living in the ex-evacuation zone set within a 20-km radius from FNPP [21]. In this review, we describe the first results of DNA damage monitoring in the cattle leashed in the ex-evacuation zone and the current situation of radiation biodosimeters.

12.2 Current Status of Biomarkers for Radiation Dose Assessment

The cellular response after irradiation is considered to correlate with radiation dose; the higher the dose is, the greater the biological effect is. In other words, to correctly understand the cellular response after radiation exposure and to predict the biological influence of radiation, biomarkers for accurate estimation of radiation dose are needed. Carcinogenic risk is one of the most concerned adverse effects of radiation exposure. Since radiation-induced DNA damage is one of the causes of cancer, it is meaningful to know how much DNA damage is induced by radiation. Currently, a method of detecting chromosomal aberrations is generally used as an evaluation index of radiation exposure risk (see Chap. 20) [22]. Among chromosomal aberrations, the background level of dicentric chromosome (DC) is almost none, and there are almost no individual differences [22, 23], so the dicentric chromosome assay (DCA) is a standard protocol of the International Organization for Standardization (ISO) as a biodosimeter that can clearly evaluate the effect of radiation exposure [24]. However, in order to prepare metaphase spreads from the collected blood sample, it is problematic that in addition to the necessity of inducing blood leukocytes division and a culture period of several days, dose can be evaluated only if it is in a relatively high-dose range [22, 23]. In addition, although many automated analysis software for detecting chromosomal aberrations including DCs are currently being developed, experienced persons are necessary for final judgment [22, 23, 25]. As the assay is labor-intensive and time-consuming, the DCA is not adequate as a dose evaluation method in large-scale exposure but as an exposure dose evaluation method for a person who is at risk of high-dose radiation exposure at the time of emergency [26–28]. On the other hand, detection of the DNA DSB level by γ-H2AX does not require cell culture [29]. Detection of DNA DSBs by γ-H2AX in peripheral blood lymphocytes (PBLs) derived from patients who have received CT scans or CT angiography has been performed, which has shown the correlation of the DNA DSB levels with exposure doses in the low-dose (LD) range of 1 mGy [29–33]. In addition, dose evaluation experiments after whole-body radiation exposure using γ-H2AX have been performed using animal models [18, 20]. Bonner's group reported the monitoring of DNA DSBs in PBLs by the γ-H2AX assay of *rhesus macaque* after whole-body irradiation of 0, 1, 3.5, 6.5, and 8.5 Gy [18]. Correlations of the dose-dependent linear DNA DSB levels were detected in all dose ranges up to 4 days after irradiation [18]. Recently, they reported the result of analysis of the DNA DSB level in PBLs after whole-body irradiation using γ-H2AX using swine model, and similarly to the previous report, linear relationship between radiation dose and γ-H2AX level was found 3 days after irradiation [20]. These data demonstrate that it is possible to estimate the exposure dose by DNA DSBs monitoring using the γ-H2AX assay. It should also be emphasized that the γ-H2AX assay is not limited by animal species [5]. The amino acid sequence of H2AX and the phosphorylation of serine residues at the time of DNA DSB induction are highly conserved in almost all eukaryotes.

12.3   Detection of DNA DSBs In Vivo in Farm Animals by γ-H2AX

After the occurrence of the FNPP accident, the monitoring of radionuclides such as cesium-134 (^{134}Cs) and ^{137}Cs has been carried out from a relatively early stage [34–40]. In contrast, dose evaluation for the assessment of biological impacts to the ecosystem was not carried out comprehensively. As mentioned above, several

Fig. 12.1 γ-H2AX foci per peripheral blood lymphocytes from cattle living in control area and the ex-evacuation zone. The box plot of γ-H2AX foci per cell was shown. (Data modified from Ref. [21])

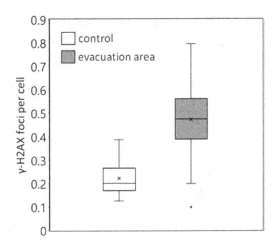

papers have been published, reporting that the radiation exposure dose assessment by the γ-H2AX assay can be conveniently performed for a wide range of animal species. Therefore, the γ-H2AX assay was performed on biospecimens derived from cattle in the ex-evacuation zone, and we monitored DNA damage level in vivo [21]. Briefly, peripheral blood samples obtained from cattle euthanized on site were shipped on ice conditions, and lymphocyte separation and detection of γ-H2AX were performed according to previously reported methods [6, 7, 21]. The γ-H2AX assay can be performed with the same antibody and the same experimental conditions even if the animal species is different. γ-H2AX detection was carried out in bovine cells under the same conditions as human cells [21]. As a result, a significantly higher level of γ-H2AX was detected compared to the control cohorts not affected by the FNPP accident, indicating that DNA DSBs were strongly induced in the cattle in the ex-evacuation zone (Fig. 12.1). Since there is a correlation between exposure dose and the DNA DSB level in vivo, the exposure dose was calculated based on the detected γ-H2AX level, and it corresponds to the radiation exposure equivalent to 20 mGy [21]. However, there was no correlation between estimated exposure dose and γ-H2AX level that is presumed from the radioactive Cs concentration remaining in the body of the same individual and/or ambient dose in the area animals were captured [21, 41]. It is the most important factor that DNA damage is repaired with time. Generally, after irradiation, the level of γ-H2AX reaches the maximum level from 30 min to 1 h after irradiation and decreases with DNA damage repair [5, 20, 42], which is restored to the level before irradiation within 24 h at the cell level [20, 42]. In other words, DNA double-strand breaks that occurred more than 72 hours ago have already been repaired and are not detectable by the γ-H2AX assay. Interestingly, as shown in Fig. 12.2, cells with many foci and cells with few foci are mixed. Although the γ-H2AX focus decreases with DNA damage repair, its rate of decrease is almost uniform, and γ-H2AX focus number shows the Poisson distribution, and there is no extreme heterogeneity [7, 20]. Repair process detected by disappearance of γ-H2AX foci occurs homogeneously if DNA DSB

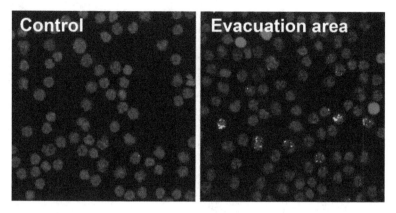

Fig. 12.2 Representative images of γ-H2AX immunostaining of peripheral blood lymphocytes from cattle living in control area and ex-evacuation zone. Green, γ-H2AX; red, DNA stained by propidium iodide

occurs simultaneously to all cells as in acute radiation exposure, but in the case of chronic exposure, DNA damage induction and damage repair might occur chronically and partly. Although the minimal DNA DSB level detected in our study is at the average about 20 mGy, more than 80% of the total cells were γ-H2AX negative, animals were exposed to chronic low-dose-rate (LDR) radiation, and the DNA DSBs induced more than 72 h before analysis could not be detected; the actual exposure dose might be higher. In fact, the γ-H2AX average value of only γ-H2AX focus positive cells was around 2.8 foci per cell (fpc), which was equivalent to radiation exposure to about 200 mGy [21]. In this study, we monitored the DNA DSB level in cattle in the ex-evacuation zone using the γ-H2AX assay and detected a high level of DNA DSBs compared to control animals [21]. Although it is difficult to evaluate accumulated dose of chronic radiation exposure by the γ-H2AX assay, our data first show biological effects of very complex radiation exposure environment such as internal and external chronic LDR exposure directly.

12.4 Perspective for Future Biodosimetry

In our previous study, to understand biological effects of the FNPP accident, we monitored DNA damage level in farm animals that were unleashed within the ex-evacuation zone of the FNPP accident and reported that higher levels of DNA DSBs were induced compared with non-affected control [21]. This indicates that DNA DSB induction by chronic LDR radiation exposure might be detected with γ-H2AX. In recent years, it has been reported that DNA DSB levels in inhabitants of high natural background radiation (HNBR) area can be detected by the γ-H2AX assay [43–45]. Although there was no significant difference, PBL derived from the HNBR population had trend to higher DNA DSB levels compared to PBLs derived from residents in other areas. Interestingly, DNA damage repair capacity in PBLs derived from the HNBR population was elevated, suggesting the existence of adaptive response by chronic LDR radiation [43, 45]. Taken together, monitoring of DNA DSBs by γ-H2AX may not be suitable for the assessment of cumulative dose of chronic LDR radiation, but is very valuable for evaluating biological effects of chronic LDR radiation exposure.

Research on biological effects by radiation exposure has been conducted based on experiments using animal models and epidemiological investigations of atomic bomb survivors of Hiroshima and Nagasaki (Hibakusha) [22, 46, 47]. Although it is evident in middle to high dose that the DNA damage level and the risk of carcinogenesis increase along with the increase of radiation dose, there is still no clear answer on biological effects below 100 mGy. Long-term and large-scale research is necessary to understand the biological effect of LDR radiation because it takes time until the effect becomes evident. Recently, the importance of tissue microenvironment change due to radiation exposure in radiation-induced cancer has been pointed out [48]. For example, the rate of radiation-induced cell differentiation is several hundred to several thousand times higher than the frequency of radiation-induced mutations [49]. This indicates that radiation dose-dependent DNA damage might not merely induce tumorigenesis of cells but tissue microenvironmental changes. In fact, there are considerable evidences showing that radiation-induced inflammation and epigenetic changes promote carcinogenesis [48, 50, 51]. Thus, to understand the biological effect of radiation exposure, it is essential not merely to investigate the number of DNA damage and the mutation frequency but also to analyze microenvironmental changes in organisms for a long time using various biological markers. We need to continue the long-term and comprehensive analysis of the ecosystem around FNPP to understand correctly the effect of persistent LDR radiation.

References

1. Rogakou EP, Pilch DR, Orr AH et al (1998) DNA double-stranded breaks induce histone H2AX phosphorylation on serine 139. J Biol Chem 273:5858–5868
2. Rogakou EP, Boon C, Redon C et al (1999) Megabase chromatin domains involved in DNA double-strand breaks in vivo. J Cell Biol 146:905–916
3. Paull TT, Rogakou EP, Yamazaki V et al (2000) A critical role for histone H2AX in recruitment of repair factors to nuclear foci after DNA damage. Curr Biol 10:886–895
4. Celeste A, Petersen S, Romanienko PJ et al (2002) Genomic instability in mice lacking histone H2AX. Science 296:922–927. https://doi.org/10.1126/science.1069398
5. Bonner WM, Redon CE, Dickey JS et al (2008) GammaH2AX and cancer. Nat Rev Cancer 8:957–967. https://doi.org/10.1038/nrc2523
6. Nakamura A, Sedelnikova OA, Redon C et al (2006) Techniques for gamma-H2AX detection. Methods Enzymol 409:236–250. https://doi.org/10.1016/s0076-6879(05)09014-2
7. Redon CE, Nakamura AJ, Sordet O et al (2011) Gamma-H2AX detection in peripheral blood lymphocytes, splenocytes, bone marrow, xenografts, and skin. Methods Mol Biol 682:249–270. https://doi.org/10.1007/978-1-60327-409-8_18

8. Sedelnikova OA, Rogakou EP, Panyutin IG et al (2002) Quantitative detection of (125)IdU-induced DNA double-strand breaks with gamma-H2AX antibody. Radiat Res 158:486–492

9. Rothkamm K, Lobrich M (2003) Evidence for a lack of DNA double-strand break repair in human cells exposed to very low x-ray doses. Proc Natl Acad Sci U S A 100:5057–5062. https://doi.org/10.1073/pnas.0830918100

10. Redon CE, Nakamura AJ, Martin OA et al (2011) Recent developments in the use of gamma-H2AX as a quantitative DNA double-strand break biomarker. Aging 3:168–174. https://doi.org/10.18632/aging.100284

11. Qvarnstrom OF, Simonsson M, Johansson KA et al (2004) DNA double strand break quantification in skin biopsies. Radiotherapy Oncol: J Eur Soc Ther Radiol Oncol 72:311–317. https://doi.org/10.1016/j.radonc.2004.07.009

12. Sak A, Grehl S, Erichsen P et al (2007) Gamma-H2AX foci formation in peripheral blood lymphocytes of tumor patients after local radiotherapy to different sites of the body: dependence on the dose-distribution, irradiated site and time from start of treatment. Int J Radiat Biol 83:639–652. https://doi.org/10.1080/09553000701596118

13. Sak A, Stuschke M (2010) Use of gammaH2AX and other biomarkers of double-strand breaks during radiotherapy. Semin Radiat Oncol 20:223–231. https://doi.org/10.1016/j.semradonc.2010.05.004

14. Redon CE, Nakamura AJ, Zhang YW et al (2010) Histone gammaH2AX and poly(ADP-ribose) as clinical pharmacodynamic biomarkers. Clin Cancer Res 16:4532–4542. https://doi.org/10.1158/1078-0432.CCR-10-0523

15. Appleman LJ, Balasubramaniam S, Parise RA et al (2015) A phase i study of DMS612, a novel bifunctional alkylating agent. Clin Cancer Res 21:721–729. https://doi.org/10.1158/1078-0432.ccr-14-1333

16. Thomas A, Redon CE, Sciuto L et al (2017) Phase I study of ATR inhibitor M6620 in combination with Topotecan in patients with advanced solid tumors. J Clin Oncol: Off J Am Soc Clin Oncol: Jco2017766915. https://doi.org/10.1200/jco.2017.76.6915

17. Balasubramaniam S, Redon CE, Peer CJ et al (2018) Phase I trial of belinostat with cisplatin and etoposide in advanced solid tumors, with a focus on neuroendocrine and small cell cancers of the lung. Anti-Cancer Drugs. https://doi.org/10.1097/cad.0000000000000596

18. Redon CE, Nakamura AJ, Gouliaeva K et al (2010) The use of gamma-H2AX as a biodosimeter for total-body radiation exposure in non-human primates. PLoS One 5:e15544. https://doi.org/10.1371/journal.pone.0015544

19. Horn S, Barnard S, Rothkamm K (2011) Gamma-H2AX-based dose estimation for whole and partial body radiation exposure. PLoS One 6:e25113. https://doi.org/10.1371/journal.pone.0025113

20. Moroni M, Maeda D, Whitnall MH et al (2013) Evaluation of the gamma-H2AX assay for radiation biodosimetry in a swine model. Int J Mol Sci 14:14119–14135. https://doi.org/10.3390/ijms140714119

21. Nakamura AJ, Suzuki M, Redon CE et al (2017) The causal relationship between DNA damage induction in bovine lymphocytes and the Fukushima nuclear power plant accident. Radiat Res 187:630–636. https://doi.org/10.1667/rr14630.1

22. Bender MA, Awa AA, Brooks AL et al (1988) Current status of cytogenetic procedures to detect and quantify previous exposures to radiation. Mutat Res 196:103–159

23. Romm H, Wilkins RC, Coleman CN et al (2011) Biological dosimetry by the triage dicentric chromosome assay: potential implications for treatment of acute radiation syndrome in radiological mass casualties. Radiat Res 175:397–404. https://doi.org/10.1667/rr2321.1

24. ISO19238 (2014) Radiological protection – Performance criteria for service laboratories performing biological dosimetry by cytogenetics

25. Wilkins RC, Romm H, Kao TC et al (2008) Interlaboratory comparison of the dicentric chromosome assay for radiation biodosimetry in mass casualty events. Radiat Res 169:551–560. https://doi.org/10.1667/rr1272.1

26. Hayata I, Kanda R, Minamihisamatsu M et al (2001) Cytogenetical dose estimation for 3 severely exposed patients in the JCO criticality accident in Tokai-mura. J Radiat Res 42(Suppl):S149–S155

27. Wojcik A, Gregoire E, Hayata I et al (2004) Cytogenetic damage in lymphocytes for the purpose of dose reconstruction: a review of three recent radiation accidents. Cytogenet Genome Res 104:200–205. https://doi.org/10.1159/000077489

28. Suto Y, Hirai M, Akiyama M et al (2013) Biodosimetry of restoration workers for the Tokyo electric power company (TEPCO) Fukushima Daiichi nuclear power station accident. Health Phys 105:366–373. https://doi.org/10.1097/HP.0b013e3182995e42

29. Lobrich M, Rief N, Kuhne M et al (2005) In vivo formation and repair of DNA double-strand breaks after computed tomography examinations. Proc Natl Acad Sci U S A 102:8984–8989. https://doi.org/10.1073/pnas.0501895102

30. Rothkamm K, Balroop S, Shekhdar J et al (2007) Leukocyte DNA damage after multi-detector row CT: a quantitative biomarker of low-level radiation exposure. Radiology 242:244–251. https://doi.org/10.1148/radiol.2421060171

31. Kuefner MA, Grudzenski S, Schwab SA et al (2009) DNA double-strand breaks and their repair in blood lymphocytes of patients undergoing angiographic procedures. Investig Radiol 44:440–446. https://doi.org/10.1097/RLI.0b013e3181a654a5

32. Kuefner MA, Grudzenski S, Hamann J et al (2010) Effect of CT scan protocols on x-ray-induced DNA double-strand breaks in blood lymphocytes of patients undergoing coronary CT angiography. Eur Radiol 20:2917–2924. https://doi.org/10.1007/s00330-010-1873-9

33. Brand M, Sommer M, Achenbach S et al (2012) X-ray induced DNA double-strand breaks in coronary CT angiography: comparison of sequential, low-pitch helical and high-pitch helical data acquisition. Eur J Radiol 81:e357–e362. https://doi.org/10.1016/j.ejrad.2011.11.027

34. Yasunari TJ, Stohl A, Hayano RS et al (2011) Cesium-137 deposition and contamination of Japanese soils due to the Fukushima nuclear accident. Proc Natl Acad Sci U S A 108:19530–19534. https://doi.org/10.1073/pnas.1112058108

35. Endo S, Kimura S, Takatsuji T et al (2012) Measurement of soil contamination by radionuclides due to the Fukushima Dai-ichi Nuclear Power Plant accident and associated estimated cumulative external dose estimation. J Environ Radioact 111:18–27. https://doi.org/10.1016/j.jenvrad.2011.11.006

36. Yoshida N, Kanda J (2012) Geochemistry. Tracking the Fukushima radionuclides. Science 336:1115–1116. https://doi.org/10.1126/science.1219493

37. Tazoe H, Hosoda M, Sorimachi A et al (2012) Radioactive pollution from Fukushima Daiichi nuclear power plant in the terrestrial environment. Radiat Prot Dosim 152:198–203. https://doi.org/10.1093/rpd/ncs222

38. Kuroda K, Kagawa A, Tonosaki M (2013) Radiocesium concentrations in the bark, sapwood and heartwood of three tree species collected at Fukushima forests half a year after the Fukushima Dai-ichi nuclear accident. J Environ Radioact 122:37–42. https://doi.org/10.1016/j.jenvrad.2013.02.019

39. Torii T, Sugita T, Okada CE et al (2013) Enhanced analysis methods to derive the spatial distribution of 131I deposition on the ground by airborne surveys at an early stage after the Fukushima Daiichi nuclear power plant accident. Health Phys 105:192–200. https://doi.org/10.1097/HP.0b013e318294444e

40. Terashima I, Shiyomi M, Fukuda H (2014) (134)Cs and (137)Cs levels in a grassland, 32 km northwest of the Fukushima 1 Nuclear Power Plant, measured for two seasons after the fallout. J Plant Res 127:43–50. https://doi.org/10.1007/s10265-013-0608-9

41. Urushihara Y, Kawasumi K, Endo S et al (2016) Analysis of plasma protein concentrations and enzyme activities in cattle within the ex-evacuation zone of the Fukushima Daiichi nuclear plant accident. PLoS One 11:e0155069. https://doi.org/10.1371/journal.pone.0155069

42. Redon CE, Dickey JS, Bonner WM et al (2009) Gamma-H2AX as a biomarker of DNA damage induced by ionizing radiation in human peripheral blood lymphocytes and artificial skin. Adv Space Res: Off J Comm Space Res 43:1171–1178. https://doi.org/10.1016/j.asr.2008.10.011

43. Jain V, Kumar PR, Koya PK et al (2016) Lack of increased DNA double-strand breaks in peripheral blood mononuclear cells of individuals from high level natural radiation areas of Kerala coast in India. Mutat Res 788:50–57. https://doi.org/10.1016/j.mrfmmm.2016.03.002

44. Hasan Basri IK, Yusuf D, Rahardjo T et al (2017) Study of gamma-H2AX as DNA double strand break biomarker in resident living in high natural radiation area of Mamuju, West Sulawesi. J Environ Radioact 171:212–216. https://doi.org/10.1016/j.jenvrad.2017.02.012

45. Jain V, Saini D, Kumar PRV et al (2017) Efficient repair of DNA double strand breaks in individuals from high level natural radiation areas of Kerala coast, South-West India. Mutat Res 806:39–50. https://doi.org/10.1016/j.mrfmmm.2017.09.003

46. Shimizu Y, Kodama K, Nishi N et al (2010) Radiation exposure and circulatory disease risk: Hiroshima and Nagasaki atomic bomb survivor data, 1950-2003. BMJ (Clinical research ed) 340:b5349. https://doi.org/10.1136/bmj.b5349

47. Ozasa K, Shimizu Y, Suyama A et al (2012) Studies of the mortality of atomic bomb survivors, report 14, 1950-2003: an overview of cancer and noncancer diseases. Radiat Res 177:229–243

48. Barcellos-Hoff MH, Park C, Wright EG (2005) Radiation and the microenvironment - tumorigenesis and therapy. Nat Rev Cancer 5:867–875. https://doi.org/10.1038/nrc1735

49. Watanabe M, Suzuki N, Sawada S et al (1984) Repair of lethal, mutagenic and transforming damage induced by X-rays in golden hamster embryo cells. Carcinogenesis 5:1293–1299

50. Aypar U, Morgan WF, Baulch JE (2011) Radiation-induced epigenetic alterations after low and high LET irradiations. Mutat Res 707:24–33. https://doi.org/10.1016/j.mrfmmm.2010.12.003

51. Morioka T, Miyoshi-Imamura T, Blyth BJ et al (2015) Ionizing radiation, inflammation, and their interactions in colon carcinogenesis in Mlh1-deficient mice. Cancer Sci 106:217–226. https://doi.org/10.1111/cas.12591

Part IV
Fukushima Accident and
Dose Estimation

13

A Study of Cattle using Electron Spin Resonance (ESR) Tooth Dosimetry

Kazuhiko Inoue, Ichiro Yamaguchi, and Masahiro Natsuhori

Abstract To validate radiation dose of cattle affected by the Fukushima Daiichi Nuclear Power Plant (FNPP) accident, we applied electron spin resonance (ESR) tooth dosimetry. Teeth were collected from cattle that had stayed continuously after the accident in Okuma Town within the ex-evacuation zone of the FNPP accident. Radiation exposure to cattle attributed to the FNPP accident was confirmed retrospectively by X-band ESR tooth dosimetry, which was almost consistent with the estimated radiation dose from airborne and individual cattle, whereas positive radiation-induced signals (RIS) were not detectable in any sample by nondestructive measurement using L-band ESR tooth dosimetry. Although ESR tooth dosimetry reflects total radiation doses of affected animals, the uncertainty of measurement was relatively large. Therefore, in order to accurately measure the additional radiation dose from the nuclear accident, it is necessary to clarify possible causes of the uncertainty. Making continuous improvements, X-band ESR tooth dosimetry for animals in the ex-evacuation zone is ongoing.

Keywords ESR tooth dosimetry · L-band · X-band · Cattle · Fukushima Daiichi Nuclear Power Plant · Nuclear disaster

K. Inoue (✉)
Department of Translational Research, Tsurumi Dental University,
Yokohama, Japan
e-mail: qb6k-inue@asahi-net.or.jp

I. Yamaguchi
Department of Environmental Health, National Institute of Public Health,
Wako, Saitama, Japan

M. Natsuhori
School of Veterinary Medicine, Kitasato University, Towada, Aomori, Japan

13.1 Introduction

Electron spin resonance (ESR) tooth dosimetry is an established method of evaluating radiation dose by measuring stable unpaired electrons produced by radiation exposure in a tooth [1, 2]. Ionizing radiation generates radicals proportionate to absorbed dose depending on radiation quality and energy. These radicals react rapidly and disappear in the most part of our body due to water, but in organized matrices without water, they can persist indefinitely. This phenomenon has been recognized in bones and teeth and has been shown to be a promising method for retrospective dosimetry, based on free radicals being stabilized in the hydroxyl apatite matrix in tooth enamel (Fig. 13.1). It has been confirmed that mean detection limit is 205 mGy, ranging from 56 to 649 mGy at the fourth International Comparison [3]. Using extracted teeth, X-band ESR tooth dosimetry has been applied to retrospective dose assessment of atomic bomb survivors of Hiroshima and Nagasaki (Hibakusha) [4, 5], the Chernobyl Nuclear Power Plant (CNPP) accident [6], nuclear bomb tests [7], and radiological accidents [8–10]. An "ideal" ESR spectrum with some parameters is shown in Fig. 13.2. The peak-to-peak amplitude "A" is the first derivative of the resonance microwave absorption. The peak-to-peak linewidth, "ΔB," is related to the modulation frequency that is important for controlling the signal-to-noise ratio. The resonance field "Br" indicates the natural resonance frequency of the radical species due to the Zeeman effect. The dose obtained with ESR dosimetry is well correlated with the standard method such as hematology-based

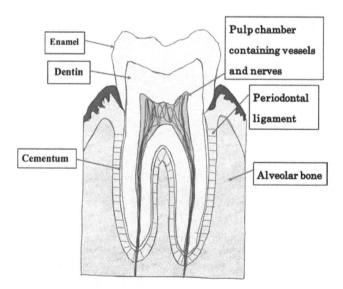

Fig. 13.1 Anatomy of a tooth and locations of tooth tissues

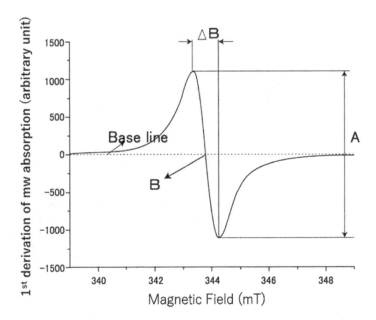

Fig. 13.2 "Ideal" ESR spectrum with illustrations of some its parameters. (Original: International Atomic Energy Agency. Use of electron paramagnetic resonance dosimetry with tooth enamel for retrospective dose assessment. IAEA Vienna IAEA-TECDOC-1331 (2002). https://www-pub. iaea.org/MTCD/Publications/PDF/te_1331_web.pdf)

assays [11]. Those dosimetry studies are carried out at conventional ESR frequencies called X-band of the electromagnetic spectrum (e.g., 9 GHz) so that the method is called X-band ESR tooth dosimetry. This X-band ESR has high sensitivity for dose estimation using the extracted or deciduous teeth. ESR spectra of tooth enamel after exposure to different doses are shown in Fig. 13.3.

With a radiation emergency in mind, an L-band ESR apparatus has been developed for in situ measurement as triage using teeth in the mouth as indices [12]. This method is available for retrospective dose assessment and for triage when workers are involved in an accident that caused radiation exposure such as a nuclear disaster or a large-scale radiation exposure accident [13].

ESR tooth dosimetry is a method of evaluating the accumulated radiation dose and is thought to be applicable for animals affected by the FNPP accident, which is crucial for studying the biological effect of persistent low-dose-rate (LDR) radiation [14–16]. The FNPP accident affected animals and several environmental study groups started research activities on livestock. We have been analyzing the influence of radiation exposure on affected cattle since September 2012. We have carried out dose estimation using information of environmental monitoring and radioactivity deposition density on the ground. We have collected 50 cattle at Okuma Farm in Okuma Town, Fukushima Prefecture (Fig. 13.4). Research of the radiation effect on large animals is still a challenging topic. To obtain scientific results, it is essential to

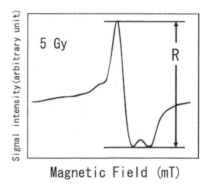

Fig. 13.3 ESR spectra of tooth enamel after 5Gy irradiation. The dosimetric component of the spectrum after irradiation with 5 Gy is shown in this chart. R is the peak-to-peak amplitude used for ESR dose reconstruction. (Original: International Atomic Energy Agency. Use of electron paramagnetic resonance dosimetry with tooth enamel for retrospective dose assessment. IAEA Vienna IAEA-TECDOC-1331 (2002). https://www-pub.iaea.org/MTCD/Publications/PDF/te_1331_web.pdf)

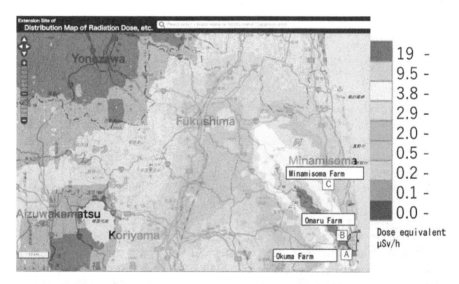

Fig. 13.4 Radiation dose-rate map obtained by airborne monitoring results and the location of Okuma Farm as of May 31, 2012. This map was quoted from the website provided by Nuclear Regulation Authority, Japan. Dose rate shows μSv/h as a dose equivalent rate including background radiation. (http://ramap.jaea.go.jp/map/)

estimate radiation dose accurately for discussing the dose response. In order to validate ESR tooth dosimetry, we estimated dose of cattle affected by the FNPP accident using ESR and compared the dose with that estimated using monitoring tendency data.

13.2 Material and Methods

13.2.1 Sample Collection

Teeth were collected from cattle that had continued to stay in Okuma Farm, that is located in the ex-evacuation zone of the FNPP accident (Fig. 13.4). Dissection of cattle was performed on May 14, 2017. Table 13.1 shows information on cattle whose radiation dose was measured in this study.

13.2.2 Nondestructive Measurement Using L-Band ESR Spectroscopy

For the comparison of the L-band ESR response among different energy photons, we used two standard materials. One was a standard tooth irradiated by the cesium-137 (^{137}Cs) source provided from the ESR Center for the Study of Viable Systems at Dartmouth, NH, USA. The other was a tooth sample exposed to 20 Gy of X-rays in an X-ray irradiator (MBR-1505R2, Hitachi Medical, Tokyo, Japan) at 150 kV and 1 mA, with 0.1-mm copper plus 0.3-mm aluminum filtering. This X-ray irradiation gives a response about 4 times as large as γ-rays from ^{137}Cs.

In order to adjust for variations in the measured radiation-induced signal (RIS) amplitude that result from instrumental instability and/or external environmental factors, the ratio of differentiated voltage as a radiation-induced signal to the differentiated voltage of 4-oxo-2,2,6,6,-tetramethylpiperidine-d16–1-^{15} N-1-oxyl (^{15}N-PDT) for each measurement was calculated and normalized to the same ratio in the standard tooth exposed to 20 Gy of 150 kV X-ray as described above.

After extracting from the jawbone portion of each cattle, a maxillary incisor tooth was washed with tap water. Then with a design knife, as much soft tissue as possible was removed and the labial side was used for the measurement surface by L-band ESR tooth dosimetry. The ESR spectrum was acquired using standard measurement parameters such as 20 scans, including a scan range of 2.5 mT, a scan time of 3 sec, and a modulation amplitude of 0.4 mT [17]. This process was repeated 3 times. A plastic tube containing a solution of ^{15}N-PDT was fixed in a position in close proximity to the surface loop and used as a reference of positive standard. Spectra from each of the collected datasets were analyzed using non-linear least-squares fitting to estimate the peak-to-peak signal amplitudes of the radiation-induced signals and of ^{15}N-PDT. RIS and signals of ^{15}N-PDT were then averaged to provide the mean amplitude for each tooth and at each dose.

Cattle	Ear tag ID	Sex	Birth date
A-1	12416–04378	Female	2006/12/26
A-2	12425-47537	Female	2007/9/2
A-3	08597–08639	Female	2012/6/1

Table 13.1 Basic characteristics of each cattle and estimated radiation dose

13.2.3 Destructive Measurement by X-Band ESR Spectroscopy

The enamel portion on the labial side of the cattle teeth was taken out by a nipper about 3 mm in size, taking care not to contain dentin, and was used as a sample for X-band ESR spectroscopy after grinding in an agate mortar to a size of 1 mm in diameter so as to enter a sample tube. ESR spectra were acquired using parameters of 40 scans, including a scan range of 5 mT, a scan time of 30 sec and a modulation amplitude of 0.2 mT power of 2 mW, and a time constant of 0.03 s. This process was repeated for a total of five datasets at each dose.

13.2.4 Measurement of Radioactivity Concentration

Radioactive concentrations of Cs and strontium-90 (^{90}Sr) in the molar of Cattle A-3 were measured. For radioactive Cs, the ashed sample was analyzed by a germanium semiconductor detector (CFG-SV-76, ORTEC, TN, USA). Sr-90 was extracted by the ion exchange method and analyzed by a low-background beta-ray measuring device after achieving radioactive equilibrium with yttrium-90 (Y-90). In principle, measurement was performed for 3,600 sec, and the radioactivity concentration was calculated.

13.2.5 Data and Statistical Analysis

Data analysis was carried out using code developed by Ivannikov [18]. Statistical analysis was performed using R3.3.1 [19]. Radiation dose of each sample was determined by the additive dose-response method assuming a linear dose-response of the ESR signals to additional irradiated doses of each sample. This method employs step-by-step irradiation to each sample additionally. Absorbed dose is estimated by a linear regression analysis [2].

13.3 Results

13.3.1 Dose Estimation by Dose-Rate Monitoring Using Survey Meters

Cumulative radiation dose based on ambient monitoring using an air chamber survey meter during 6 years since March 2011 was estimated to be 270 mGy [20].

13.3.2   Dose Estimation by Radiation Dose Monitoring Using Individual Dosimeters

Radiation dose measured by an individual dosimeter attached to representative cattle from December 2013 to December 1, 2017 was 57 mGy (Fig. 13.5). Considering weathering, total dose from the day the FNPP accident happened to the start of monitoring was calculated from deposition density of iodine-131 (^{131}I), tellurium-132 (^{132}Te), ^{132}I, ^{134}Cs and ^{137}Cs on the ground immediately after the accident [21]. Estimated total dose of Cattle A-1 and A-2 from March 2011 was 160 mGy. Similarly total dose of Cattle A-2 calculated was 67 mGy after adjusting deposition density of radionuclides to its birth month.

13.3.3   Dose Estimation by ESR Tooth Dosimetry

Positive RIS were not detectable in any cattle sample examined by nondestructive measurement using L-band ESR dosimetry (Fig. 13.6). For comparison, an exposed human tooth is shown in Fig. 13.7. In this spectrum, RIS were detected between 0.7 and 1.1 mT.

A representative X-band ESR tooth spectroscopy of Cattle A-1 is shown in Fig. 13.8. The additive-dose-response for radiation dose estimation is shown in Fig. 13.9. The dose of this sample is the value of negative x-intercept value, 67 mGy. The uncertainty of dose estimation was calculated using the Monte Carlo method by picking up randomly generated parameters 100 times, adapting the model obtained by the additive-dose method. Estimated 95% confidence interval was between 1.3

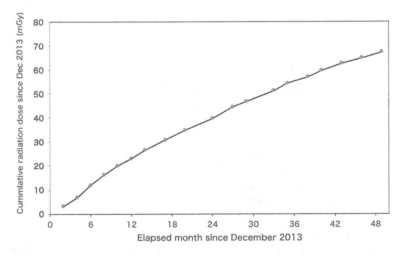

Fig. 13.5 Cumulative radiation doses to cattle monitored by using glass dosimeters since December 2013

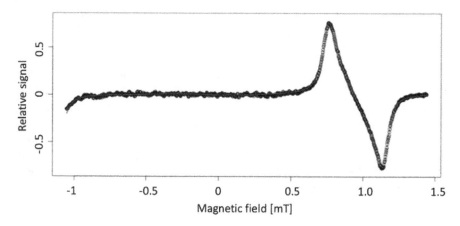

Fig. 13.6 Representative ESR spectrum of a sample of incisors from Cattle A-2 using L-band spectrometry. Measurement condition: 20 scans, scan range of 2.5 mT, a scan time of 30 sec, a modulation amplitude of 0.4 mT. The signal around 0.8–1.2 mT shows a positive reference control of ^{15}N-PDT

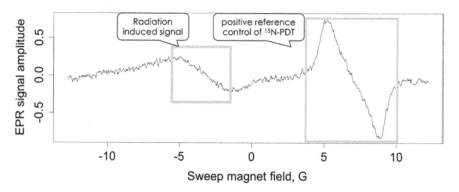

Fig. 13.7 Example of ex vivo ESR spectrum of human tooth using the L-band. This tooth is exposed to 1Gy by X-ray. Measurement condition: 20 scans, scan range of 2.5 mT, a scan time of 3 sec, a modulation amplitude of 0.4 mT. The signal from a positive control is observed around 0.6–0.9 mT. Radiation induced signal is observable at 0.5–1 mT

and 119 mGy. Total dose estimated by X-band ESR tooth dosimetry was compared with other methods, that is, calculated from ambient dose-rate and from individual monitors after adding the initial estimated dose before December 2013 (Table 13.2). Both estimations using ground-deposited radioactive Cs and individual monitor were in good agreement. Although the dose range was large, doses estimated by X-band ESR tooth dosimetry of Cattle A-1 and A-3 were not much different from those calculated by other two methods. In Cattle A-2, the dose assessed by X-band ESR tooth dosimetry was eight times the dose assessed by other two methods.

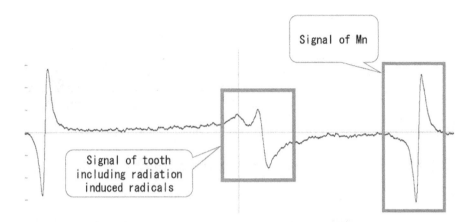

Fig. 13.8 Example of spectrum obtained by X-band ESR tooth dosimetry. In the middle, tooth signals are shown including radiation-induced signal. At both sides, there are standard signals from a Mn marker. The right side is third signal of Mn and the left side is fourth signal of Mn

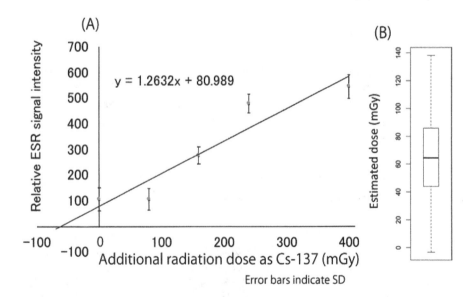

Fig. 13.9 Estimated radiation dose of a sample from Cattle A-1 by the additive-dose method. (**a**) Irradiated dose was converted from X-ray to ^{137}Cs equivalent considering the energy dependency of this ESR tooth dosimetry. Dose is the value of negative x-intercept value, 67 mGy. (**b**) The uncertainty of this dose estimation. The top of the box represents the third quartile and the bottom of the box represents the first quartile. The upper whisker and the lower beard show 1.5 interquartile range. Outliers are plotted as individual points. In this measurement, estimated dose was 67 ± 30 mGy showing lower value for this cattle

Table 13.2 Estimated radiation dose [mGy]

Cattle	Based on initial ground deposition and ambient dose monitoring	Based on individual dose assessment[a]	ESR range, median
A-1	158	160	37–731, 255
A-2	158	160	1120–1230, 1180
A-3	65	67	114–221, 138

[a]individual doses are averaged among monitored cattle

13.3.4   Radioactivity Concentration in the Molar Tooth of Cattle A-3

Radionuclides and their concentrations detected in the ashed molar tooth of Cattle A-3 were ^{90}Sr (46 ± 2.3 Bq kg), ^{134}Cs (60.7 ± 9.9 Bq/kg), and ^{137}Cs (60.6 ± 1.7 Bq/kg).

13.4   Discussion

13.4.1   Comparisons of Estimated Radiation Dose

Total dose using X-band ESR tooth dosimetry was 37–1230 mGy. These values are consistent with the estimated radiation dose using monitoring and individual dosimeters, considering that Cattle A-2, who showed had the highest exposure dose, was moved from the higher dose area to Okuma Farm during the 6 years.

13.4.2   Limitations of ESR Tooth Dosimetry in Regard to Variance and Energy Dependency

This study tried to measure the actual dose of cattle affected by the FNPP accident using both X-band and L-band ESR tooth dosimetries. Due to its low sensitivity, L-band ESR tooth dosimetry revealed to be not practical for dose assessments associated with the FNPP accident and that the X-band dosimetry can be available with some improvements. Dose of each sample was measured by the additive-dose method. Measurements were performed on each enamel sample individually, taking into account the sample properties and mass. This is because, even with the same tooth, the degree of stable radical generation is affected by the chemical structure of the enamel sample. Response characteristics of thermoluminescence dosimeter (TLD) depend on the mass energy-absorption coefficient of photons and the mass stopping power of charged particles. In dose measurements, energy dependence of ESR dosimetry is relatively similar to TLD. When measuring with animal samples by the additive dose method, the dose response may vary from sample to sample.

Therefore, it is more desirable to measure separately for each sample. In fact, there is significant variation in dose response of ESR dosimetry among samples of the same boar, especially in older boars [22].

13.4.3   *Effect of External Exposure Due to β-Particles from Deposited Radionuclides*

Assuming that the deposited density of ^{137}Cs was 1 MBq/m^2, external dose-rate of the skin in contact with the ground was estimated to be around 0.14 mGy/h using coefficient of ICRU report 56 [23]. Assuming 8 h a day of contact with the ground, the total skin dose from β-particles for 6 years after the FNPP accident was 1.9 Gy with consideration of weathering. It is a large value in the dose calculation and cannot be ignored. However, this estimated dose was not detectable by ESR tooth dosimetry, indicating that β-particle exposure to the teeth is negligible unless getting a tooth close to the land.

13.4.4   *Effect of Internal Exposure*

Other concern was a potential impact of internal exposure by radioactive substances inside a tooth and other organs. However, self-absorption dose, that is, radiation dose to a tooth from radionuclides in the same tooth, is limited. It was because of the low concentration of the total radionuclides in tooth, and taking the radioactive strontium as an example, the tooth dose was as low as 2 mGy.

13.4.5   *Limitations of ESR Tooth Dosimetry in Regard to the Variance of Background Signals*

The background signal in enamel could be induced by ultraviolet from sunlight [24, 25] or mechanical stimulation during sample preparation [26]. Therefore, further confirmation on the added radiation dose due to the nuclear accident is needed to clarify the attribution of possible causes.

13.5   Conclusion

Radiation dose measured by ESR tooth dosimetry reflects the radiation history of each subject. In this study, the dose measured by ESR tooth dosimetry was consistent with the monitored radiation dose. These indicate that teeth can be used to assess radiation dose from a radiation accident. However, L-band ESR dosimetry was not sensitive enough to detect the radiation exposure to cattle in Okuma Town.

Acknowledgments The authors are sincerely grateful to the farmers who have donated their precious cattle. Authors also thank Prof. Shin Toyoda and Prof. Harold Swartz for their kind encouragement and valuable suggestions.

This work was in part supported by the Industrial Disease Clinical Research Grants, the Ministry of Health, Labour and Welfare (Grant No. 150803-02), and Grant-in-Aid, Japan Society for the Promotion of Science (JSPS; KAKENHI Grant No. 15 K11435).

References

1. Brady JM, Aarestad NO, Swartz HM (1968) In vivo dosimetry by electron spin resonance spectroscopy. Health Phys 15(1):43–47
2. IAEA (2002) TECDOC-1331. Use of electron paramagnetic resonance dosimetry with tooth enamel for retrospective dose assessment
3. Fattibenea P, Wieser A, Adolfsson E et al (2011) The 4th international comparison on EPR dosimetry with tooth enamel: Part 1: report on the results. Radiat Meas 46:765–771
4. Nakamura N, Hirai Y, Kodama Y (2012) Gamma-ray and neutron dosimetry by EPR and AMS, using tooth enamel from atomic-bomb survivors: a mini review. Radiat Prot Dosim 149:79–83
5. Ikeya M, Miki T, Kai A, Hoshi M (1986) ESR dosimetry of A-bomb radiation using tooth enamel and granite rocks. Radiat Prot Dosim 17:181–184
6. Chumak VV (2012) The Chernobyl experience in the area of retrospective dosimetry. J Radiol Prot 32:N59–N63
7. Romanyukha A, Schauer DA, Malikov YK (2006) Analysis of current assessments and perspectives of ESR tooth dosimetry for radiation dose reconstruction of the population residing near the Semipalatinsk nuclear test site. J Radiat Res 47(Suppl A):A55–A60
8. Tolstykh EI, Degteva MO, Kozheurov VP et al (2000) Strontium metabolism in teeth and enamel dose assessment: analysis of the Techa River data. Radiat Environ Biophys 39:161–171
9. Trompiera F, Battaglinia P, Bey E (2008) EPR dosimetry in recent radiation accident cases. Radioprotection 43:184
10. Shiraishi K, Iwasaki M, Miyazawa C et al (2002) Dose estimation by ESR on tooth enamel from two workers exposed to radiation due to the JCO accident. J Radiat Res 43:331–335
11. Nakamura N, Miyazawa C, Sawada S et al (1998) A close correlation between electron spin resonance (ESR) dosimetry from tooth enamel and cytogenetic dosimetry from lymphocytes of Hiroshima atomic-bomb survivors. Int J Radiat Biol 73:619–627
12. Miyake M, Liu KJ, Walczak TM, Swartz HM (2000) In vivo EPR dosimetry of accidental exposures to radiation: experimental results indicating the feasibility of practical use in human subjects. Appl Radiat Isot 52:1031–1038
13. Swartz HM, Flood AB, Williams BB et al (2014) Comparison of the needs for biodosimetry for large-scale radiation events for military versus civilian populations. Health Phys 106:755–763
14. Jiao L, Liu ZC, Ding YQ et al (2014) Comparison study of tooth enamel ESR spectra of cows, goats and humans. J Radiat Res 55:1101–1106
15. Toyoda S, Romanyukha A, Hino Y, Itano S et al (2007) Effect of chemical treatment on ESR dosimetry of cow teeth: application to samples from the southern Urals. Radiat Meas 42:1178–1180
16. Zdravkova M, Galleza B, Debuyst R (2005) A comparative in vivo and in vitro L-band EPR study of irradiated rat incisors. Radiat Meas 39:143–148

17. Yamaguchi I, Sato H, Kawamura H et al L-band EPR tooth dosimetry for heavy ion irradiation. Radiat Prot Dosim 172:81–86
18. Ivannikov A, Sanin D, Nalapko M et al (2010) Dental enamel EPR dosimetry: comparative testing of the spectra processing methods for determination of radiation-induced signal amplitude. Health Phys 98:345–351
19. R Development Core Team (2005) R: a language and environment for statistical computing. R Foundation for Statistical Computing, Vienna, Austria. ISBN 3-900051-07-0
20. Sasak Ji SH, Deguchi Y et al (2019) Decreased blood cell counts were not observed in cattle living in the "difficult-to-return zone" of the Fukushima nuclear accident. Anim Sci J 90:128–134
21. Natsuhori M, Kojima T, Sato I (2017) Radioactive contamination profiles of soil, external and internal exposure to Japanese Black Cattle after Fukushima Daiichi Nuclear Plant Accident. Jpn J Large Anim Clin 8:143–147
22. Harshman A, Toyoda S, Johnson T (2018) Suitability of Japanese wild boar tooth enamel for use as an electron spin resonance dosimeter. Radiat Measure 116:46–50
23. ICRU (1997) Report 56. Dosimetry of External Beta Rays for Radiation Protection
24. Nakamura N, Katanic JN, Miyazawa C (1998) Contamination from possible solar light exposures in ESR dosimetry using human tooth enamel. J Radiat Res 39:185–191
25. Sholom S, Desrosiers M, Chumak V et al (2010) UV effects in tooth enamel and their possible application in EPR dosimetry with front teeth. Health Phys 98:360–368
26. Aragno D, Fattibene P, Onori S (2001) Mechanically induced EPR signals in tooth enamel. Appl Radiat Isot 55:375–382

14

Japanese Macaques: Cumulative Dose due to External and Internal Exposures

Satoru Endo, Kenichi Ishii, Masatoshi Suzuki, Tsuyoshi Kajimoto, Kenichi Tanaka, and Manabu Fukumoto

Abstract Cumulative dose of external and internal exposures following the Fukushima Daiichi Nuclear Power Plant (FNPP) accident was estimated for Japanese macaques (*Macaca fuscata*) around FNPP. Conversion factors for Japanese macaques modeled as ellipsoids were estimated for external exposure from contaminated ground and internal exposure uniformly distributed in the body. Conversion factors for seven radionuclides, namely, tellurium 129 (^{129}Te), ^{129m}Te, iodine-131 (^{131}I), ^{132}Te, ^{132}I, cesium-134 (^{134}Cs) and ^{137}Cs were calculated using the PHITS code. The estimated factors for the seven radionuclides were consistent with those in ICRP Publication 108, using an effective radius for comparison. The external, internal and total exposures for 13 macaques in Namie Town were estimated by applying the calculated factors. The estimated cumulative exposures for the periods from the accident occurred to the sampling date, ranged from 0.26 to 1.6 Gy. The average exposure was 0.64 Gy in averaged over the 11 sampled macaques except for the 2 macaques which might be born after the FNPP accident.

Keyword Fukushima Daiichi Nuclear Power Plant accident · Dose rate conversion factor · External exposure · Internal exposure · Japanese macaques

S. Endo (✉) · K. Ishii · T. Kajimoto · K. Tanaka
Quantum Energy Applications, Graduate School of Engineering, Hiroshima University, Higashi-Hiroshima, Hiroshima, Japan
e-mail: endos@hiroshima-u.ac.jp

M. Suzuki
Institute for Disaster Reconstruction and Regeneration Research, Tohoku University, Sendai, Japan

M. Fukumoto
Institute of Development, Aging and Cancer, Tohoku University, Sendai, Japan

School of Medicine, Tokyo Medical University, Tokyo, Japan

14.1 Introduction

The nuclear accident at the Fukushima Daiichi Nuclear Power Plant (FNPP) was triggered by an enormous earthquake and associated tsunami (the Great East Japan Earthquake) on March 11, 2011. A large quantity of radioactive nuclides were released, resulting in the severe contamination of a wide area of the southern Tohoku region to the northern Kanto region. The main depositions occurred on March 15 and March 20–21, 2011 [1, 2].

Since the accident, many studies have examined the effect of radiation on animals, insects and plants [3–8]. However, cumulative exposure in animals is quite difficult to estimate because it is challenging to account for habitat and behavioral properties. Estimates of exposure could be improved using a contamination map combined with habitat information. Dose rate and its temporal changes can be estimated for target animals in an area from the contamination map.

The National Nuclear Security Administration (NNSA) performed a rapid survey of radiation and contamination on March 17–19, 2011 using the Aerial Measuring System. These data were used for the initial NNSA March 22, 2011, report and are available for independent analysis [9]. Additionally, the Ministry of Education, Culture, Sports, Science and Technology, Japan (MEXT), conducted a contamination study using a 2-km mesh from June to August 2011 [10]. This study started 3 months after the main deposition on March 15, 2011. The radionuclides with short half-lives already decayed away; however, ^{134}Cs and ^{137}Cs remained over a wide area around FNPP.

Studies of the biological effect of radiation following the FNPP accident were initiated by Tohoku University as a part of a Comprehensive Dose Evaluation Project Concerning Animals Affected by The Fukushima Daiichi Nuclear Power Plant Accident and a Nuclear Energy Science and Technology and Human Resource Development Project by MEXT using Japanese macaques (*Macaca fuscata*) sampled in the Fukushima area. Macaques, like humans, belong to the order, Primates. Therefore, the biological effect on macaques might provide important information that can be applied to humans. To estimate exposure, the International Commission of Radiological Protection (ICRP) summarized conversion factors for reference animals under internal exposure to uniformly distributed radionuclides and external exposure to contaminated ground with bodies simulated as ellipsoids in ICRP Publication 108 [11]. The factors are available for 75 radionuclides from ICRP Publication 108. However, the Japanese macaque is not one of the ICRP reference animals. Therefore, we estimated the appropriate conversion factors for the species using the Monte Carlo technique.

14.2   Materials and Methods

14.2.1   *^{137}Cs Contamination Map*

NNSA [9] and MEXT data [10] for ^{137}Cs contamination were combined and inter-polated by the Universal Kriging method using System for Automated Geoscientific Analyses–Geographic Information System (SAGA–GIS) software [12]. A vario-gram was generated, which describes the variance of the difference between ^{137}Cs concentrations at two locations fitted by a quartic function, using SAGA-GIS [10]. The reproducibility of the variogram was 95% within 70 km from FNPP. The inter-polated values were stored in 1,814 × 1,854 meshes for latitudes from 36.51611 to 38.32911 and longitudes from 139.30506 to 141.15896. The interpolated map is shown in Fig. 14.1. Using this map, ^{137}Cs contamination at the coordinates where macaques were sampled can be calculated.

14.2.2   *Ellipsoid Model*

In ICRP Publication 108, the reference animals in terrestrial conditions are modeled as various ellipsoids [11], as listed in Table 14.1. Therefore, Japanese macaques were also modeled as ellipsoids with a density of 1 g/cm^3 according to ICRP Publication 108 assumption. Japanese macaques were categorized into three groups

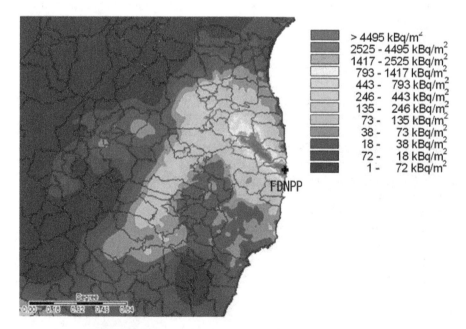

Fig. 14.1 Kriging interpolated map for the combined NNSA and MEXT data

Table 14.1 Body parameters for the reference animals in ICRP Publication 108 [11] and the calculated effective radius

Reference animal	Body mass (kg)	Body shape proportions (ellipsoid)	Effective radius (cm)
Deer	245	$1 \times 0.4620 \times 0.4620$	38.7
Rat	0.314	$1 \times 0.3000 \times 0.2500$	4.21
Duck egg	5.03×10^{-2}	$1 \times 0.6670 \times 0.6670$	2.29
Duck	1.26	$1 \times 0.3330 \times 0.2670$	6.69
Frog egg	5.24×10^{-4}	$1 \times 1 \times 1$	0.500
Frog	0.0314	$1 \times 0.375 \times 0.313$	1.96
Bee colony	28.3	$1 \times 0.500 \times 0.500$	18.8
Bee	5.89×10^{-4}	$1 \times 0.375 \times 0.375$	0.520
Earthworm	5.24×10^{-3}	$1 \times 0.100 \times 0.100$	1.08

Table 14.2 Japanese macaque models for the walk and sit postures based on physical measurements

Axis (cm)	Walk model			Sit model		
	<5 kg (S)	5–10 kg (M)	>10 kg (L)	<5 kg (S)	5–10 kg (M)	>10 kg (L)
Major	15	25	30	15	25	30
First minor	6.2	9.7	12	6.2	9.7	12
Second minor	4.1	6.4	8	4.1	6.4	8
Height	20	30	40	–	–	–

according to the body size, namely, small (S), middle (M) and large (L), with the weight of less than 5 kg, 5–10 kg and greater than 10 kg, corresponding to the sampled macaques. The ellipsoid dimensions were based on the physical measurements of 52 macaques obtained by Tohoku University and the values are listed in Table 14.2. Two types of postures, the walk and sit postures, were considered and are also listed in Table 14.2.

14.2.3 Conversion Factor Calculation

The conversion factors for internal exposure to uniformly distributed radionuclides in an ellipsoid and external exposure to uniformly contaminated ground from 0 to 0.5 cm were calculated using PHITS code [13]. The calculation geometry of air occupying a $10 \times 10 \times 10$ m area and 50-cm-deep ground soil was taken into account. The ellipsoid macaque was set at the center of the ground surface. To minimize calculation time, the four side boundaries were set to a mirror condition. Gamma-rays and β-particles were separately calculated for seven radionuclides, namely, ^{129}Te, ^{129m}Te, ^{131}I, ^{132}Te, ^{132}I, ^{134}Cs and ^{137}Cs. The γ-ray energies and emission rates for the radionuclides were obtained from the National Nuclear Data Center [14]. Beta-particle energy spectra were obtained from the literature [15, 16], and the internal conversion electrons for each radionuclide were obtained from the website of the National Nuclear Data Center [14]. The energy deposition in the ellipsoids was calculated for each γ-ray and each electron source. In addition, to consider the

Fig. 14.2 Calculation
geometry for PHITS

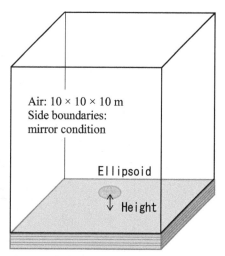

Air: $10 \times 10 \times 10$ m
Side boundaries:
mirror condition

Ellipsoid

↓ Height

Soil layer: 0-0.5 cm, 0.5-1 cm, 1-2 cm, 2–5

cm, 5–10 cm, 10–15 cm, and 15–20 cm.

depth profile of radionuclides, the deposition energy in the walk model for body size
L was also calculated for sources uniformly distributed in six soil layers, namely,
0.5–1 cm, 1–2 cm, 2–5 cm, 5–10 cm, 10–15 cm and 15–20 cm (Fig. 14.2).

The external exposure conversion factors were derived by weighing each deposition energy for the depth profile of the radionuclide. The factor D_i^{ex} ((mGy/day)/
(Bq/m^2)) of the ith radionuclide from depth layer d_k is expressed by

$$D_i^{ex}(d_k) = \sum_{j=\gamma,\beta} \frac{E_{ij}(d_k)}{m} I_j S \cdot 3600 \cdot 24 \cdot 1000, \qquad (14.1)$$

where subscript j indicates emitted γ-rays or β-particles, I_j is the emission rate for
emitted γ-rays or β-particles, m is the mass of the macaque-simulating ellipsoid, E
is the deposition energy in the ellipsoid by emitted γ-rays and β-particles and S is
the calculated area expressed in m^2. In the case of soil surface contamination, similar to the ICRP Publication 108 assumption, the depth layer d_l is only 0–0.5 cm. To
estimate the factor for the depth profile of radionuclides, the fraction of radionuclides as a function of depth, $f(d_k)$, is introduced and summed for seven layers:

$$D_i^{ex} = \sum_7^{k=1} f_1(d_k) \cdot D_i^{ex}(d_k) \qquad (14.2)$$

The internal exposure conversion factor D_i^{in} ((mGy/d)/(Bq/kg)) for each radionuclide i was calculated using each deposition energy as follows:

$$D_i^{in} = \sum_{j=\gamma,\beta} \frac{E_{ij}}{m} I_j m \cdot 3600 \cdot 24 \cdot 1000. \qquad (14.3)$$

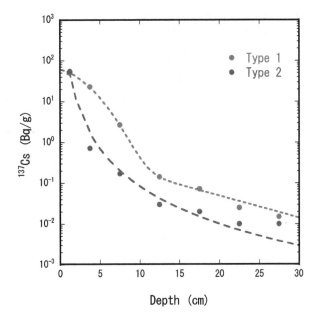

Fig. 14.3 Tested depth profiles. Type 1, farmland; Type 2, normal land

14.2.4   Depth Profile

There is a lack of depth profile data, except for radioactive Cs. The depth profiles of ^{137}Cs radioactivity have been measured at various locations [17–20]. The relaxation mass depths of ^{137}Cs have been reported as time-dependent values of 1.1–1.5 g/cm^2 for the period from December 2012 to December 2013 [17]; 0.9 g/cm^2 for soil with a low clay content [18]; 1.56 g/cm^2, on average, for paddy soil [19]; and 1.04, 1.38, 1.05 and 1.29 g/cm^2 in April 2011, October 2011, December 2011 and March 2012, respectively [20]. The conversion factor for the ^{137}Cs depth-profiled soil was calculated using the relaxation mass depths from 0.5 to 2.7 according to Eq. (14.2). Additionally, the typical depth profiles for our measurement, as shown in Fig. 14.3, were tested.

14.3   Results and Discussion

14.3.1   Conversion Factor

The calculated conversion factors for the surface layer for seven radionuclides are listed in Table 14.3. The external exposure results for the walk model and the internal exposure results were compared with those in ICRP Publication 108 against the effective radius and are shown in Fig. 14.4. The calculated factor and those of ICRP Publication 108 agree within 20%. Therefore, if we allow for 20% uncertainty, the conversion factor in ICRP Publication 108 for the effective radius can be used.

Table 14.3 Estimated conversion factors for the external exposure rates of the walk and sit model for surface contamination and the internal exposure rate

| Radionuclide | Body group | Conversion factor [(mGy/d)/(Bq/m²)] | | Conversion factor [(mGy/d)/(Bq/kg)] |
		External, walk model	External, sit model	Internal
¹³⁴Cs	L	9.50×10^{-5}	1.00×10^{-4}	7.63×10^{-3}
	M	1.02×10^{-4}	1.07×10^{-4}	6.72×10^{-3}
	S	1.17×10^{-4}	1.16×10^{-4}	5.14×10^{-3}
¹³⁷Cs	L	3.53×10^{-5}	3.77×10^{-5}	4.91×10^{-3}
	M	3.81×10^{-5}	3.93×10^{-5}	4.58×10^{-3}
	S	4.32×10^{-5}	4.31×10^{-5}	4.01×10^{-3}
¹³²I	L	1.17×10^{-4}	1.22×10^{-4}	9.87×10^{-3}
	M	1.24×10^{-4}	1.33×10^{-4}	9.27×10^{-3}
	S	1.46×10^{-4}	1.44×10^{-4}	8.25×10^{-3}
¹³¹I	L	2.23×10^{-5}	2.21×10^{-5}	5.02×10^{-3}
	M	2.40×10^{-5}	2.35×10^{-5}	4.60×10^{-3}
	S	2.75×10^{-5}	2.60×10^{-5}	3.88×10^{-3}
¹³²Te	L	1.18×10^{-5}	9.95×10^{-6}	3.57×10^{-3}
	M	1.29×10^{-5}	1.08×10^{-5}	3.15×10^{-3}
	S	1.45×10^{-5}	1.18×10^{-5}	2.43×10^{-3}
¹²⁹ᵐTe	L	3.26×10^{-6}	3.09×10^{-6}	7.86×10^{-3}
	M	4.20×10^{-6}	3.68×10^{-6}	7.83×10^{-3}
	S	5.81×10^{-6}	5.20×10^{-6}	7.77×10^{-3}
¹²⁹Te	L	6.34×10^{-6}	5.55×10^{-6}	6.48×10^{-3}
	M	7.48×10^{-6}	6.50×10^{-6}	6.43×10^{-3}
	S	1.18×10^{-5}	9.58×10^{-6}	6.34×10^{-3}

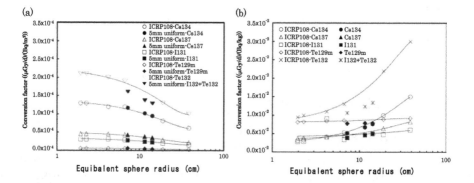

Fig. 14.4 (a) External exposure for the walk model and (b) internal exposure for the Japanese macaque. Values were compared with those in ICRP Publication 108 against the effective radius

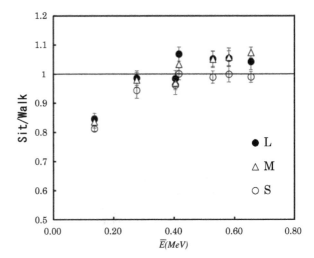

Fig. 14.5 Conversion factor ratio of the walk-to-sit model as a function of $\bar{E}$

The conversion factor ratio of external exposure for the walk to the sit model is plotted as a function of $\bar{E}$, which is defined as the sum of mean γ-ray and β-particle energies, in Fig. 14.5. The ratio is consistent with unity within statistical uncertainty, except for the lowest $\bar{E}$ of ^{132}Te. The ratio for ^{132}Te, which emits low-energy γ-rays and β-particles, is approximately 0.83. This is due to the shielding effect by an individual's own body. Therefore, except for low-energy β-particles and γ-rays, the conversion factors for both the walk model and sit model are approximately equal.

14.3.2   Conversion Factor for Depth-Profiled Radionuclides

The conversion factors for the seven depth-profiled radionuclides with a relaxation depth from 0.6 to 2.6 and measured depth profiles of ^{134}Cs and ^{137}Cs in Iitate Village, Fukushima Prefecture, in comparison with the 0–0.5 cm uniform factor are shown in Fig. 14.6a, b, respectively. The relaxation depths of ^{137}Cs radioactivity from previous studies [17–20] are shown in light red and ranged from 0.9 to 1.5 g/cm². A soil self-shielding factor introduced as a ratio of the conversion factor for the depth profiles to that for the uniformly distributed factor in surface soil characterized by the relaxation depth is summarized in Table 14.4. The soil self-shielding factors for ^{134}Cs and ^{137}Cs were 0.74 and 0.71, respectively. The shielding factor for Type 1 for the slowest measured depth profiles for farmland in Fig. 14.6b was less than that for a relaxation depth of 2.6 g/cm². However, the shielding factor for Type 2 for the typical depth profile for undisturbed land is consistent with those for the relaxation depth of 0.9–1.5 g/cm². The variation within the range of 0.9–1.5 g/cm² corresponds to the variation in the conversion factor for ^{134}Cs or ^{137}Cs, which was less than 8%.

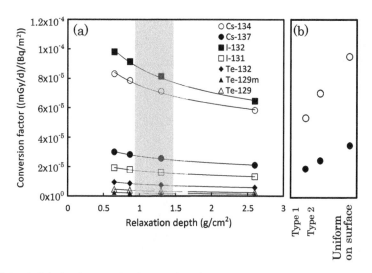

Fig. 14.6 (**a**) Calculated conversion factors of walk model for seven radionuclides of relaxation depth and (**b**) typical measured depth profiles

Table 14.4 Soil self-shielding factor for seven radionuclides

Radionuclide	Soil shielding factor
^{134}Cs	0.74
^{137}Cs	0.71
^{132}I	0.68
^{131}I	0.71
^{132}Te	0.62
^{129m}Te	0.54
^{129}Te	0.52

14.3.3 Example of External Exposure Estimation

To consider cumulative exposure to soil contamination, temporal changes in the air dose from radionuclides deposited on the ground are good references. The air dose rate from seven radionuclides from radioactivity concentrations was estimated around FNPP in 2011 [9, 10]. The air dose rate changes were estimated for several locations. For example, the estimated dose rate change in Iitate Village assuming only the physical half-life [2] is shown in Fig. 14.7. Integrating over time, the cumulative air dose can be estimated. The cumulative air dose normalized by the dose at 200 years after deposition and the radionuclide contribution as a function of time are shown in Fig. 14.8a, b, respectively. The sum of the ^{134}Cs and ^{137}Cs contributions to the cumulative air dose at 1 year after deposition or later exceeds 90%. In the sampled macaques, exposure is expected to be the same as the contributions to the air dose. Accordingly, the external exposure for some samples after 1 year following the FNPP accident can be estimated by only radioactive ^{134}Cs and ^{137}Cs, neglecting the 10% contribution of external exposure to other radionuclides.

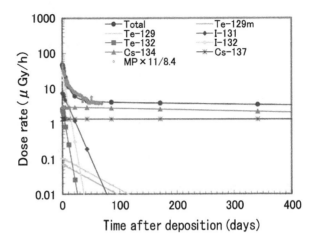

Fig. 14.7 Estimated dose rate change in Iitate Village assuming only the physical half-life

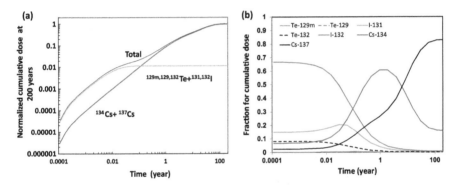

Fig. 14.8 (a) Cumulative dose in Japanese macaques against time since deposition normalized by the dose at 200 years; (b) radionuclide contribution as a function of time

Using the ^{134}Cs and ^{137}Cs conversion factors for the walk model, the external, the internal and the total exposures were calculated for the sampled macaques at Namie Town, Fukushima Prefecture, in 2013. Only ^{137}Cs concentration and ^{134}Cs concentration, which is the same as ^{137}Cs concentration in March 2011, were used in this estimation. The ground concentration of ^{137}Cs (interpolated values in Fig. 14.1) and the ^{137}Cs concentration in macaque bodies measured at Tohoku University [21] were used in this calculation. To account for the effect of the depth profile, the soil self-shielding factor of the depth profile for a relaxation depth of 1.3 g/cm^2 was assumed to be 0.74 for ^{134}Cs and 0.71 for ^{137}Cs. The macaques were caught in 2013 or later; therefore, only exposure to ^{134}Cs and ^{137}Cs was taken into account. Additionally, the ratio of the accumulated concentration of ^{134}Cs to ^{137}Cs was assumed to be unity. The estimated total exposures ranged from 0.26 to 1.6 Gy, as shown in Table 14.5. Two sampled macaques showed the body weight less than 2kg. There is a possibility that these 2 macaques were born after the FNPP accident. The average exposure was 0.64 Gy in averaged over the 11 sampled macaques except for the 2 macaques.

Table 14.5 Summary of results for a sample of Japanese macaques in Namie Town. $\dot{D}$ and D show the exposure rates and cumulative exposure. $t = 0$ indicates the values at the time of deposition

ID	Latitude	Longitude	Sampling date	Days from accident	Body length (cm)	Weight (kg)	^{137}Cs accumulation in soil (kBq/ m²)	^{137}Cs in the body (Bq/kg)	^{137}Cs $\dot{D}_{in}$ (t = 0) (μGy/ day)	^{134}Cs $\dot{D}_{in}$ (t = 0) (μGy/ day)	D_{in} (mGy)	^{137}Cs $\dot{D}_{ex}$ (t = 0) × 0.71 (μGy/day)	^{134}Cs $\dot{D}_{ex}$ (t = 0) × 0.74 (μGy/day)	D_{ex} (mGy)	Total dose (mGy)
7	37.55924	140.7524	2013/3/6	726	80	–	2105	35,834	184.3	286.1	282	52.8	148.0	116	398
64	37.54273	140.8626	2013/10/28	962	50	5.8	3026	26,238	127.9	187.5	241	81.9	228.0	224	465
217	37.54179	140.8619	2013/11/8	973	45	4.7	3151	10,654	47.3	60.5	122	96.6	272.9	373	495
221	37.49008	140.9414	2015/7/10	1582	66.5	15.6	1467	6921	37.8	58.6	110	36.8	103.1	146	255
233	37.50656	140.9104	2015/9/28	1662	60	12	2349	4730	25.9	40.3	78	58.9	165.1	240	318
252	37.49175	140.9326	2015/12/15	1740	60	11.9	2139	11,760	64.9	100.7	202	53.6	150.3	227	429
404	37.47923	140.9886	2016/3/24	1840	49	–	6657	33,742	189.1	293.4	640	16.7	46.8	76	716
405	37.54214	140.8606	2016/11/22	2083	52	9.5	3412	65,323	341.5	500.8	1131	92.4	257.1	419	1550
406^a	37.54214	140.8606	2016/11/22	2083	32	2	3412	31,550	144.5	184.9	454	104.6	295.5	478	932
413	37.54121	140.8628	2016/12/2	2093	60	13	3115	65,581	367.8	570.8	1251	78.1	218.9	356	1608
414^a	37.51081	140.8954	2016/12/2	2093	31	1.4	2119	4791	22.0	28.1	69	65.0	183.5	298	367
477	37.47565	140.9402	2017/12/8	2464	47	9.5	2610	2012	10.8	15.8	40	70.7	196.7	353	393
478	37.47231	140.9399	2017/12/12	2468	46	8.5	2993	1816	9.7	14.3	36	81.0	225.6	405	441

^{a}There is a possibility that the 2 macaques were born after the FNPP accident

14.4 Conclusions

The cumulative external and internal exposures attributed to the FNPP accident were estimated for Japanese macaques around FNPP. Conversion factors for Japanese macaques, modeled as ellipsoids, for external exposure from contaminated ground and internal exposure uniformly distributed in the body were calculated. The estimated conversion factors for seven radionuclides, namely, ^{129}Te, ^{129m}Te, ^{131}I, ^{132}Te, ^{132}I, ^{134}Cs and ^{137}Cs, agreed within 20% with those in ICRP Publication 108 using an effective radius.

The conversion factors for depth-profiled ^{134}Cs and ^{137}Cs characterized by the relaxation depth were also estimated. The conversion factors are expressed by the factor for a uniform distribution at the 0–0.5 cm surface multiplied by a soil self-shielding factor of 0.74 for ^{134}Cs and 0.71 for ^{137}Cs and with a relaxation depth of 1.3 g/cm^2.

The external, the internal and the total exposures for 13 sampled macaques were estimated using the calculated conversion factors. The estimated cumulative exposures for the periods from the accident occurred to the sampling date, ranged from 0.26 to 1.6 Gy. The average exposure was 0.64 Gy in averaged over the 11 sampled macaques except for the 2 macaques which might be born after the FNPP accident.

Acknowledgments The authors are grateful to Prof. Chary Rangacharyulu, the University of Saskatchewan, for his advice. The radioactivity measurements were performed at the radiation facility of the Graduate School of Engineering, Hiroshima University, and at the Department of Chemistry, Faculty of Science, and Graduate School of Science, Tohoku University. The authors are grateful to Ms. Aya Kado at the radiation facility of Hiroshima University and Mr. Hiroaki Tamaki, Mr. Takumi Ono and Mr. Kazuma Koarai of Tohoku University. This work is supported by the Nuclear Energy Science and Technology and Human Resource Development Project (Through Concentrating Wisdom) (281302) from the Ministry of Education, Culture, Sports, Science and Technology of Japan.

References

1. Endo S, Kimura S, Takatsuji T et al (2012) Measurement of soil contamination by radionuclides due to Fukushima Daiichi Nuclear Power Plant accident and associated cumulative external dose estimation. J Environ Radioact 111:18–27
2. Imanaka T, Endo S, Sugai M et al (2012) Early radiation survey of the Iitate Village heavily contaminated by the Fukushima Daiichi accident, conducted on 28 and 29 March 2011. Health Phys 102:680–686
3. Fukuda T, Kino Y, Abe Y et al (2013) Distribution of artificial radionuclides in abandoned cattle in the evacuation zone of the Fukushima Daiichi nuclear power plant. PLoS One 8:e54312
4. Ochiai K, Hayama S, Nakiri S et al (2014) Low blood cell counts in wild Japanese monkeys after the Fukushima Daiichi nuclear disaster. Sci Rep 4:5793
5. Takahashi S, Inoue K, Suzuki M et al (2015) A comprehensive dose evaluation project concerning animals affected by the Fukushima Daiichi Nuclear Power Plant accident: its set-up and progress. J Radiat Res 56(S1):i36–i41
6. Hiyama A, Nohara C, Kinjo S et al (2012) The biological impacts of the Fukushima nuclear accident on the pale grass blue butterfly. Sci Rep 2:270

7. Akimoto S (2014) Morphological abnormalities in gall-forming aphids in a radiation-contaminated area near Fukushima Daiichi: selective impact of fallout? Ecol Evol. https://doi.org/10.1002/ece3.949

8. Hayama S, Nakiri S, Nakanish S et al (2013) Concentration of radiocesium in the wild Japanese monkey (*Macaca fuscata*) over the first 15 months after the Fukushima Daiichi nuclear disaster. PLoS One 8:e68530

9. Musolino SV, Clark H, McCullough T et al (2012) Environmental measurements in an emergency: this is not a drill. Health Phys 102(5):516–526

10. Minister of Education, Culture, Sports, Science and Technology (MEXT) (2011) Map of radiocesium in soil. http://www.mext.go.jp/b_menu/shingi/chousa/gijyutu/017/shiryo/_icsFiles/afieldfile/2011/09/02/1310688_2.pdf. Last accessed 29 Aug 2011

11. International Commission on Radiological Protection (ICRP) (2008) Environmental protection: the concept and use of reference animals and plants, Ann. ICRP 38 (4–6), ICRP publication 108

12. Cimmery V SAGA user guide, updated for SAGA version 2.0.5, 2007–2010. http://www.saga-gis.org/en/index.html. Last accessed 8 July 2018

13. Sato T, Niita K, Matsuda N et al (2013) Particle and heavy ion transport code system PHITS, version 2.52. J Nucl Sci Technol 50:913–923

14. National Nuclear Data Center (NNDC). NNDC data base: interactive chart of nuclides. http://www.nndc.bnl.gov/chart/. Last update: 1 February 2017. Last accessed 12 Feb 2017

15. Endo S, Tanaka K, Kajimoto T et al (2014) Estimation of β-ray dose in air and soil from Fukushima Daiichi Power Plant accident. J Radiat Res 55:476–483

16. Endo S, Kajimoto T, Tanaka K et al (2015) Mapping of cumulative β-ray dose on the ground surface surrounding the Fukushima area. J Radiat Res 56:i48–i55

17. Matsuda N, Mikami S, Shimoura S et al (2015) Depth profiles of radioactive cesium in soil using a scraper plate over a wide area surrounding the Fukushima Dai-ichi Nuclear Power Plant, Japan. J Environ Radioact 139:427–434

18. Kato H, Onda Y, Teramage M (2012) Depth distribution of ^{137}Cs, ^{134}Cs, and ^{131}I in soil profile after Fukushima Dai-ichi Nuclear Power Plant accident. J Environ Radioact 111:59–64

19. Shiozawa S, Tanoi K, Nemoto K et al (2011) Vertical concentration profiles of radioactive caesium and convective velocity in soil in a paddy field in Fukushima. Radioisotopes 60(8):323–328

20. Honda M, Matsuzaki H, Miyake Y et al (2015) Depth profile and mobility of ^{129}I and ^{137}Cs in soil originating from the Fukushima Dai-ichi Nuclear Power Plant accident. J Environ Radioact 146:35–43

21. Suzuki M. (2015) Compiled data of Japanese monkeys sampled in Fukushima Area by A comprehensive dose evaluation project concerning animals affected by the Fukushima Daiichi Nuclear Power Plant accident, Personal communications

Part V
Aftermath

15

A Study of Radioactive Cs in the Environment

Kazuhiko Ninomiya

Abstract A large amount of radionuclides were released into the environment by the Fukushima Daiichi Nuclear Power Plant (FNPP) accident. By this accident, radioactive cesium (Cs)-bearing particles were also released. Since these particles stably exist in the environment, the influence of long-term exposure to radiation is concerned if the particle is ingested in the human body. In this review, in order to evaluate the influence of radioactive Cs-bearing particles on the environment, studies on radioactive Cs-bearing particles are summarized. Radioactive Cs-bearing particles can be classified into several kinds depending on their elemental composition and radioactivity. Although the production and distribution of radioactive Cs-bearing particles have still been unclear, details of the situation inside the nuclear reactor at the accident can be investigated from the classification and investigation of their production.

Keywords Fukushima Daiichi Nuclear Power Plant · Radioactive Cs-bearing particles · Radioactivity analysis · Cs contamination inventory

15.1 Introduction

On March 11, 2011, a large magnitude 9.0 earthquake occurred in the Pacific Coast of the northeast region of Japan and an accident occurred at the Fukushima Daiichi Nuclear Power Plant (FNPP) due to the large tsunami generated by the earthquake. This accident was classified in the most serious category of nuclear accident by the International Atomic Energy Agency (level 7), and a large amount of radionuclides were released into the atmosphere and ocean [1]. In addition to measuring the air

K. Ninomiya (✉)
Department of Chemistry, Graduate School of Science, Osaka University,
Toyonaka, Osaka, Japan
e-mail: ninokazu@chem.sci.osaka-u.ac.jp

dose rate, the distribution of nuclides, such as iodine-131 (^{131}I), tellurium-129m (^{129m}Te), silver-110m (^{110m}Ag) and radioactive cesium (^{134}Cs and ^{137}Cs) deposited in soil was measured, and the initial contamination status was determined [2].

In a nuclear accident, volatilized radionuclides leak from the damaged fuel body and are released into the environment and captured by aerosols in the atmosphere. The aerosol behavior changes according to wind direction, speed and other factors at the time of the release event from the nuclear reactor, greatly affecting the transportation process [3]. Sulfuric acid is the most abundant aerosol in the atmosphere resulting in the radionuclide being transported as a sulfuric acid aerosol form. For the FNPP accident, by analyzing the contents of ionic compounds and radioactivities in the atmospheric aerosol for various particle sizes, it was determined that the amount of radioactive Cs strongly correlates with the amount of sulfuric acid aerosol [4].

On the other hand, heterogeneous spot-like contamination was reported via autoradiography analysis using imaging plates of environmental samples such as plant, soil and atmospheric filters just after the accident [5, 6]. Heterogeneous spot-like contamination was never washed out by water and was not the result of soluble sulfuric acid aerosols but arose from particles with concentrated radioactivity. Radioactive Cs is strongly adsorbed by minerals such as mica and some heterogeneous contamination was identified as soil particles [7–9]. However, Adachi et al. [10] reported in 2013 the detection of insoluble radioactive particles with spherical shapes of several micrometers in diameter from air samples collected in 2011. SEM image of the insoluble particle and its elemental composition is shown in Fig. 15.1. The particles are called variously such as radioactive Cs concentrated particles, Cs particles, insoluble particles, radiocesium-bearing microparticles (CsPs) and radioactive Cs-rich microparticles (CsMPs). The author calls the particles "radioactive Cs-bearing particles" hereafter in this article. Radioactive Cs-bearing particles are quite different from the "hot particles" found in the Chernobyl NPP accident [11] and various radioactive Cs-bearing particles have been discovered in Fukushima Prefecture and neighboring areas to date.

In this review, the author summarizes the recently reported papers on radioactive Cs-bearing particles that have been commonly detected after the FNPP accident. In addition to the properties of radioactive Cs-bearing particles, we examine their trace element analysis, distribution in the environment, formation processes and prospects for future research.

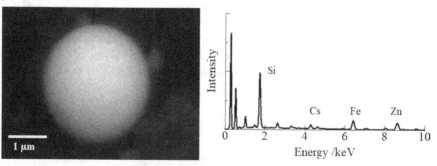

Fig. 15.1 SEM image of a radioactive Cs-bearing particle (left) and its EDS spectrum (right)

15.2 Properties of Radioactive Cs-Bearing Particles

Radioactive Cs-bearing particles were isolated and analyzed for the first time from atmospheric samples (filters) collected in Tsukuba, 170 km southwest from FNPP [10]. From the scanning electron microscope (SEM) analysis, it is confirmed that radioactive Cs-bearing particles are spherical with a diameter of 2 μm and contain Cs, iron (Fe), zinc (Zn), manganese (Mn) and chlorine (Cl). Those radioactive Cs-bearing particles also contain several Bq of ^{134}Cs and ^{137}Cs with the particle radioactivity and size, indicating that a particle contains approximately 5 wt% of radioactive Cs. The particle is primarily composed of SiO_2 though it is not clear because there are silicon (Si) background signals in the SEM analysis. They also reported radioactive Cs-bearing particles retain their shape even after washing with water. Abe et al. [12] conducted a more detailed analysis of radioactive Cs-bearing particles by synchrotron radiation X-ray (SR-μ-X-ray) analysis showing that radioactive Cs-bearing particles contain molybdenum (Mo), barium (Ba), tin (Sn) and a trace amount of uranium (U). Furthermore, the chemical state of the metals contained in radioactive Cs-bearing particles was determined by SR-μ-XANES (synchrotron radiation X-ray absorption near edge structure) analysis, and Fe, Zn, Sn and Mo are present in their oxidized states as Fe^{3+}, Zn^{2+}, Sn^{4+} and Mo^{6+}, respectively. Radioactive Cs-bearing particles can also be isolated from leaves, soil and dust in the high dose area of Fukushima Prefecture [13–15]. Sato et al. [16] performed detailed analysis of ^{134}Cs and ^{137}Cs radioactivity for radioactive Cs-bearing particles found in soil collected 20 km northwest of FNPP and reported the ^{134}Cs/^{137}Cs activity ratio just after the accident to be 1.03 ± 0.01. The authors also reported the clear linear relationship between particle size and radioactivity, that is, the specific activity is the same, indicating that radioactive Cs-bearing particles originate from the same source (Fig. 15.2).

Fig. 15.2 Relation between the size of a radioactive Cs-bearing particle and its activity obtained by Sato et al. [16]

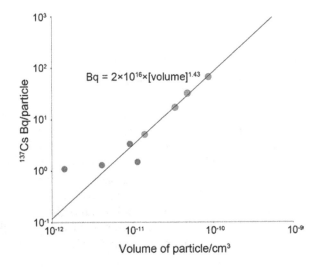

$Bq = 2 \times 10^{16} \times [volume]^{1.43}$

^{137}Cs Bq/particle

Volume of particle/cm³

Table 15.1   Characteristics of radioactive Cs-bearing particles [15]

Characteristic	Particle type	
	A	B
Size distribution	1–10 μm	40–400 μm
^{134}Cs/^{137}Cs	1.04	0.93
Other radionuclides	N/A	Antimony-125 (^{125}Sb)
Distribution	Wide	Limited (North)
Suspected emission date	March 15, 2011	March 12, 2011
Source of reactor(s)	Unit 2 or 3	Unit 1
Specific radioactivity	High	Low
Elements in common	Si, O, Zn	
Elements contained	Fe, Sn, Cl	Na, Mg, Ca, Ba

Sato et al. [15] classified 19 radioactive Cs-bearing particles collected from soil and dust samples in Fukushima Prefecture and found two types of radioactive Cs-bearing particles depending on the discovery location (Table 15.1). Both radioactive Cs-bearing particles are primarily composed of SiO_2 as a base material [17], but their detailed properties are different. The "type A" particles are similar to the particles found in Tsukuba [10, 12]. These particles are characterized by several Bq of radioactivity per particle, but the specific activity is very large due to the small particle size, and typical radioactive Cs content is approximately several wt%. The ^{134}Cs/^{137}Cs ratio was approximately 1, so the source of the particle is likely Unit 2 or Unit 3, as estimated by Nishihara et al. [18]. The particles classified as "type B" have large activities, some exceeding 20 kBq per particle. On the other hand, the specific activity is much smaller than that of type A particle because the size of the type B particle is typically several tens of μm or more. The ^{134}Cs/^{137}Cs ratio is approximately 0.9, and the source of the type B particles is likely Unit 1 [18]. In addition, some type B particles exhibit Antimony-125 (^{125}Sb) activity in addition to radioactive Cs.

15.3   Detailed Structure of Radioactive Cs-Bearing Particles by Destructive Analysis Method

Yamaguchi et al. [13] and Kogure et al. [19] sliced radioactive Cs-bearing particles into a thin film using a focused ion beam (FIB) instrument and analyzed the film by transmission electron microscopy (TEM), X-ray microanalysis with scanning TEM (STEM) and energy-dispersive X-ray spectroscopy (EDS). These analyses clearly showed that radioactive Cs-bearing particles are primarily silicate glass composed of Si and O. From the semiquantitative analysis by TEM of the entire particle, the content of SiO_2 was estimated to be 70%. It is also reported that the constituent elements are distributed almost uniformly in the radioactive Cs-bearing particle. However, the concentration of radioactive Cs outside the particles is approximately twice as high as the inside concentration, and potassium (K) and rubidium (Rb) are in the opposite distributions. The authors also reported that sulfide nanoparticles are implanted in the particles [13].

Furuki et al. [14] found that radioactive Cs-bearing particles contain fine pores 10–200 nm in size as determined by TEM analysis. In addition, they reported that crystalline nanostructures such as fine CsOH and Fe-Zn-oxide are formed inside the particles.

Imoto et al. [20] analyzed the isotope ratios of U, Cs, Ba, Rb, K and calcium (Ca) by secondary ion mass spectrometry (SIMS). They reported the $^{235}U/^{238}U$ isotope ratio to be 0.29, which is significantly higher than the natural abundance of U (0.00729). In SIMS analysis, it is difficult to distinguish between Cs and Ba isobars, but the isotopic ratio of Ba indicates that most Ba contained in radioactive Cs-bearing particles originated from the decay of radioactive Cs. The $^{87}Rb/^{85}Rb$ isotope ratio also indicated that Rb derived from fission products. On the other hand, ratios of K and Ca were consistent with their respective natural isotopic ratios. The authors also attempted to measure the amount of ^{90}Sr, but the concentration fell below the limit of detection. Ochiai et al. [21] reported that U in radioactive Cs-bearing particles is present in iron oxide nanoparticles detected by high-angle annular dark-field scanning transmission electron microscopy (HAADF-STEM).

15.4   Separating Radioactive Particles

Radioactive Cs-bearing particles are very small and contain a SiO_2 matrix. Many environmental particles have similar elemental compositions, so identification and extraction of radioactive Cs-bearing particles can be difficult. Although bulk extraction has not yet been developed, other separation methods have been developed to date, as discussed below.

Adachi et al. [10] succeeded in isolating radioactive Cs-bearing particles by fragmenting the sample. From autoradiography measurements using an imaging plate (IP), the position of radioactive Cs-bearing particles in the environmental sample was identified. They transfered the environmental sample including radioactive Cs-bearing particles to a carbon tape and separated it into small pieces. By repeating the visualization by IP and fragmentation of the carbon tape, the particle could be completely separated from other components in the environmental sample. Furuki et al. [14] developed a similar method for isolating the radioactive Cs-bearing particles from soil samples. The authors disperse the soil sample on grid paper and identify the position of radioactive Cs-bearing particles by autoradiography and then drop pure water on that position. The droplet is spreaded thinly and the position of the radioactive Cs-bearing particles is determined using IP again. By repeating this procedure, radioactive Cs-bearing particles are separated from the bulk soil sample.

Radiation measurement using a semiconductor or scintillation detector instead of autoradiography by IP is also used to separate radioactive Cs-bearing particles.

Kurihara et al. [22] placed the environmental samples including radioactive Cs-bearing particles in a vial and divided each sample into two parts after mixing. After radiation measurement, water was added to the higher activity sample and subsequently divided into two samples. By repeating this procedure approximately 30 times, the radioactive Cs-bearing particles can be completely separated from the bulk soil.

15.5   Environmental Fate of Radioactive Cs-Bearing Particles

Ikehara et al. [23] conducted an autoradiography experiment with IP to determine the total amount of radioactive particles in soil samples. The soil was divided with a 114-mesh filter and the number of radioactive particles was determined by counting the spots of radioactivity on IP. The authors identified radioactive particles (> 0.06 Bq) in surface soil collected from Okuma Town and Iitate Village and found 48–318 radioactive particles per gram of soil, and the ratio of total radioactivity of the particles to the entire soil sample was 8.53–31.8%.

Igarashi et al. [24] reported the amount of radioactive particles in environmental samples by separating radioactive Cs-bearing particles from soil samples collected in Okuma Town, Fukushima Prefecture. Initially, the number of radioactive Cs-bearing particles (with radioactivity of >1 Bq) was identified by autoradiography and was separated using Kurihara's method [22]. The radioactivity ratio was approximately 0.2%, differing from the results of Ikehara et al. [23]. This indicates that the proportion of radioactive Cs-bearing particles differs greatly depending on the sampling location.

Miura et al. [25] successfully isolated radioactive particles from suspended particles in rivers. From 2011 to 2016, suspended particles were collected by passing river water collected from the Kuchibuto River in Fukushima Prefecture through a filter. The authors isolated radioactive Cs-bearing particles found on the filter with a radioactivity more than 0.4 Bq, and the activity ratio of radioactive Cs-bearing particles to the filter was estimated to be 0–46%. The particles with weaker radioactivity (0.1–0.4 Bq) were identified via the IP analysis and total radioactivity on the filter was estimated to be 1.3–67%.

Yamaguchi et al. [26] discussed the deterioration of radioactive Cs-bearing particles in the environment and determined the weathering of the particles and heterogeneity of the constituent Cs and Rb [13, 19]. The authors concluded that these phenomena are caused by several environmental factors including dissolution by cycling of wet and dry states as well as shape change caused by weathering kinetics of silicate glass. The lifetime of the particles is predicted to be shorter than several decades.

Okumura et al. [27] reported the stability of ^{134}Cs and ^{137}Cs by heating radioactive Cs-bearing particles. The authors conducted a heating experiment on six radioactive particles and showed that the radioactivity decreased at 600 °C and was completely lost when heated to 1,000 °C. The size and characteristics of the heated particles were unchanged after heating. Upon heating, elements other than Cs, such as Cl and K, were also lost as they were likely released from the particles in a similar manner. The authors also reported the heating experiment of radioactive Cs-bearing particles with soil samples and showed that the radioactivity in radioactive Cs-bearing particles moved to soil, causing significant dilution. They concluded that radioactivity in radioactive Cs-bearing particles can be decomposed by the treatment in incinerators with a burning temperature of approximately 900 °C.

15.6   Production of Radioactive Cs-Bearing Particles

Radioactive Cs-bearing particles are not produced in the environment due to their high specific radioactivity and stability and are only produced in nuclear reactors and can be released by accidents. Determination of their production process is important to estimate the total amount of radioactivity released to the environment and to investigate the conditions in the nuclear reactor during the accident. Although the process of producing radioactive particles is unclear, several hypotheses have been proposed through detailed analysis of the produced radioactive particles.

The source of Si and O, the main components of the particle, also remains unclear. Kogure et al. [19] showed that SiO_2 was derived from concrete and fine particles of SiO_2 which were formed by the molten core concrete interaction (MCCI) after the meltdown. Satou et al. [16] indicated that SiO_2 is also contained in the heat insulating material with a lower melting point than that of concrete and is located near the nuclear reactor containment vessel. Kobata et al. [28] characterized the production of $CsFeSiO_4$ under high temperature conditions by CsOH chemisorption to stainless steel containing Si.

Kogure et al. [19] measured the diffusion coefficients of alkali metals and investigated that the diffusion coefficient of Cs is two orders of magnitude smaller than K or Rb. The different distribution of alkali metals is derived from the various rates of diffusion of volatized alkali metals from the surface of the silicate particles. The authors also fabricated glass with the same composition in an attempt to reproduce the results of radioactive Cs-bearing particles. Imoto et al. [20] detected Zn-Fe-oxide nanoparticles in radioactive Cs-bearing particles, noting that iron nanoparticles are formed under high temperature conditions by MCCI and the evaporated SiO_2 and CsOH can aggregate on nanoparticles. Ochiai et al. [21] pointed out that U is simultaneously taken into iron nanoparticles because U is found only in iron nanoparticles.

Fujita et al. [29] proposed a particle production process including the alkali melting reaction of CsOH. It is well-known that radioactive Cs evaporated from the fuel body in the form of CsOH, which is a strong base. The authors succeeded in producing particles by dissolving insulation materials or concrete mainly composed of SiO_2 into droplets of CsOH melted at 300 °C as shown in Fig. 15.3.

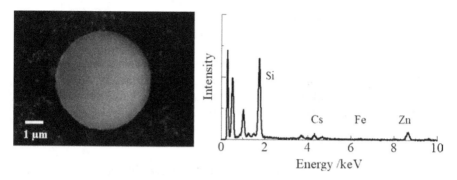

Fig. 15.3 SEM image of a reproducing radioactive Cs-bearing particle produced by the method of Fujita et al. [29] (left) and EDS spectrum of the reproducing particle (right)

15.7   Future Study Prospects

Herein, the author discussed the properties of concentrated radioactive particles found in the environment after the FNPP accident. These particles are insoluble in water and contain SiO_2 as the main component. Although the production and distribution of radioactive Cs-bearing particles remain unclear, their detailed elemental composition is reported and classified into several distinct types of particles present in the environment. Studies related to particle classification and investigation of their production can clarify the release process of radionuclides at the time of the accident.

Acknowledgment The authors thank Dr. Satou (Japan Atomic Energy Agency) for providing data on the radioactive Cs-bearing particles. This work was partially supported by the grant the Nuclear Energy Science & Technology and Human Resource Development Project (Through Concentrating Wisdom) #281302 and the Grant-in-Aid for Scientific Research A #16H01777 from the Ministry of Education, Culture, Sports, Science and Technology of Japan.

References

1. Chino M, Nakayama H, Nagai H et al (2011) Preliminary estimation of release amounts of ^{131}I and ^{137}Cs accidentally discharged from the Fukushima Daiichi Nuclear Power Plant into the atmosphere. J Nucl Sci Technol 48:1129–1134
2. Saito K, Tanihata I, Fujiwara M et al (2015) Detailed deposition density maps constructed by large-scale soil sampling for gamma-ray emitting radioactive nuclides from the Fukushima Daiichi Nuclear Power Plant accident. J Environ Radioact 139:308–319
3. Tsuruta H, Oura Y, Ebihara M et al (2014) First retrieval of hourly atmospheric radionuclides just after the Fukushima accident by analyzing filter-tapes of operational air pollution monitoring stations. Sci Rep 4:6717
4. Kaneyasu N, Ohashi H, Suzuki F et al (2012) Sulfate aerosol as a potential transport medium of radiocesium from the Fukushima nuclear accident. Environ Sci Technol 46:5720–5726

5. Tanoi K, Hashimoto K, Sakurai K et al (2011) An imaging of radioactivity and determination of Cs-134 and Cs-137 in wheat tissue grown in Fukushima. Radioisotopes 60:317–322 (in Japanese)
6. Nakanishi TM, Kobayashi NI, Tanoi K (2013) Radioactive cesium deposition on rice, wheat, peach tree and soil after nuclear accident in Fukushima. J Radioanal Nucl Chem 296:985–989
7. Kogure T, Morimoto K, Tamura K et al (2012) XRD and HRTEM evidence for fixation of cesium ions in vermiculite clay. Chem Lett 41:380–382
8. Okumura T, Tamura K, Fujii E et al (2014) Direct observation of cesium at the interlayer region in phlogopite mica. Microscopy 63:65–72
9. Motai S, Mukai H, Watanuki T et al (2016) Mineralogical characterization of radioactive particles from Fukushima soil using μ–XRD with synchrotron radiation. J Mineral Petrol Sci 111:305–312
10. Adachi K, Kajino M, Zaizen Y et al (2013) Emission of spherical cesium-bearing particles from an early stage of the Fukushima nuclear accident. Sci Rep 3:2554
11. Salbu B, Kerkling T, Oughton DH (1998) Characterization of radioactive particles in the environment. Analyst 123:843–850
12. Abe Y, Iizawa Y, Terada Y et al (2014) Detection of uranium and chemical state analysis of individual radioactive microparticles emitted from the Fukushima nuclear accident using multiple synchrotron radiation X-ray analyses. Anal Chem 86:8521–8525
13. Yamaguchi N, Mitome M, Akiyama-Hasegawa K et al (2016) Internal structure of cesium-bearing radioactive microparticles released from Fukushima nuclear power plant. Sci Rep 6:20548
14. Furuki G, Imoto J, Ochiai A et al (2017) Caesium-rich micro-particles: a window into the meltdown events at the Fukushima Daiichi Nuclear Power Plant. Sci Rep 7:42731
15. Satou Y, Sueki K, Sasa K et al (2018) Analysis of two forms of radioactive particles emitted during the early stages of the Fukushima – Nuclear Power Station accident. Geochem J 52:137–143
16. Satou Y, Sueki K, Sasa K et al (2016) First successful isolation of radioactive particles from soil near the Fukushima Daiichi Nuclear Power Plant. Anthropocene 14:71–76
17. Ono T, Iizawa Y, Abe Y et al (2017) Investigation of the chemical characteristics of individual radioactive microparticles emitted from reactor 1 by the Fukushima Daiichi Nuclear Power Plant accident by using multiple synchrotron radiation X-ray analyses. Bunseki Kagaku 66:251–261 (in Japanese)
18. Nishihara K, Iwamoto H, Suyama K (2012) Estimation of fuel compositions in Fukushima-Daiichi Nuclear Power Plant. JAEA-Data/Code 2012-018
19. Kogure T, Yamaguchi N, Segawa H et al (2016) Constituent elements and their distribution in the radioactive Cs-bearing silicate glass microparticles released from Fukushima nuclear plant. Microscopy 65:451–459
20. Imoto J, Ochiai A, Furuki G et al (2017) Isotopic signature and nano-texture of cesium-rich micro-particles: release of uranium and fission products from the Fukushima Daiichi Nuclear Power Plant. Sci Rep 7:540 9
21. Ochiai A, Imoto J, Suetake M et al (2018) Nuclear fuel fragments released to the environment with cesium-rich microparticles from the Fukushima Daiichi Nuclear Power Plant. Environ Sci Technol 52:2586–2594
22. Kurihara Y, Miura H, Higaki S et al (2017) Radioactive cesium-bearing particles in black substances, 2017 Annual Meeting of the Japan Society of Nuclear and Radiochemical Science
23. Ikehara R, Suetake M, Komiya T et al (2018) Novel method of quantifying radioactive Cesium-rich microparticles (CsMPs) in the environment from the Fukushima Daiichi Nuclear Power Plant. Environ Sci Technol 52:6390–6398
24. Igarashi J, Zhang Z, Ninomiya K et al (2018) Properties of the insoluble particles including radioactive Cs found in Okuma town, Fukushima prefecture. Proceedings of the 19th Workshop on Environmental Radioactivity. in press (in Japanese)

25. Miura H, Kurihara Y, Sakaguchi A et al (2018) Discovery of radiocesium-bearing micropar-
    ticles in river water and their influence on the solid-water distribution coefficient (Kd) of radio-
    cesium in the Kuchibuto River in Fukushima. Geochem J 52:145–154
26. Yamaguchi N, Kogure T, Mukai H et al (2018) Structures of radioactive Cs-bearing micropar-
    ticles in non-spherical forms collected in Fukushima. Geochem J 52:123–136
27. Okumura T, Yamaguchi N, Dohi T et al (2018) Loss of radioactivity in radiocesium-bearing
    microparticles emitted from Fukushima Daiichi nuclear power plant by heating. Sci Rep
    8:9707
28. Kobata M, Okane T, Nakajima K et al (2018) Chemical form analysis of reaction products
    in Cs-adsorption on stainless steel by means of HAXPES and SEM/EDX. J Nucl Mater
    498:387–394
29. Fujita N, Ninomiya K, Zhang Z et al (2016) Production simulation experiments of the insol-
    uble Cs-concentrated particles released by the FDNPP accident and their imaging on a filter.
    Proceedings of the 17th Workshop on Environmental Radioactivity, KEK Proceedings 2016-8,
    153–157

16

Insoluble Radioactive Cs Bearing Particles and their Physicochemical Properties

Masatoshi Suzuki, Kazuhiko Ninomiya, Yukihiko Satou, Keisuke Sueki, and Manabu Fukumoto

Abstract Insoluble radioactive particles have been found in the terrestrial, aquatic and aerial environments. Hot particles are well known as insoluble radioactive particles found after the nuclear tests and the accident at Chernobyl Nuclear Power Plant (CNPP). Hot particles are highly radioactive pieces and mainly composed of nuclear fuel. Insoluble radioactive particles were also found following the Fukushima Daiichi Nuclear Power Plant (FNPP) accident. The particles dissipated from FNPP are almost made of amorphous silica and condensed radioactive cesium (Cs); therefore, they are referred to as radioactive Cs-bearing particles. Radioactive Cs-bearing particles show radioactivity several orders of magnitude lower than hot particles; however, their adverse effects on human health are of great concern. This article summarizes physicochemical properties of radioactive Cs-bearing particles so far reported and discusses their biological effects.

Keywords Fukushima Daiichi Nuclear Power Plant accident · Radioactive cesium · Insoluble particles · Biological effects

M. Suzuki (✉)
Institute for Disaster Reconstruction and Regeneration Research, Tohoku University,
Sendai, Japan
e-mail: masatoshi.suzuki.c7@tohoku.ac.jp

K. Ninomiya
Department of Chemistry, Graduate School of Science, Osaka University,
Toyonaka, Osaka, Japan

Y. Satou
Collaborative Laboratories for Advanced Decommissioning Science,
Japan Atomic Energy Agency, Fukushima, Japan

K. Sueki
Center for Research in Isotopes and Environmental Dynamics, University of Tsukuba,
Ibaraki, Japan

M. Fukumoto
Institute of Development, Aging and Cancer, Tohoku University, Sendai, Japan

School of Medicine, Tokyo Medical University, Tokyo, Japan

16.1   Radioactive Particles

Nuclear tests and nuclear accidents disperse large amounts of radioactive substances into the environment [1, 2]. Following radioactive events, radioactive particles are occasionally found not only in the terrestrial and aerial but also in aquatic environments due to their insoluble nature [3, 4]. As these radioactive particles are derived from nuclear weapons or nuclear fuels, they are generally referred to as "hot particles." Japanese fishermen were exposed to the fallout from the Castle Bravo nuclear test at Bikini Atoll on March 1, 1954, which was a representative event leading to exposure to hot particles. Hot particles are also generated during normal operations at nuclear installations, and it has been reported that hot particles were inadvertently scattered early in the development of nuclear power [5]. Current concerns center around particles related to nuclear installations. Hot particles of the CNPP accident ranged from hundreds to a few micrometers in size [6]. A study conducted in Poland showed that radioactivity of hot particles from the CNPP accident ranged 1–19 kBq per particle [7]. Radioactive particles were also found after the FNPP accident initially in aerosol filters. Those particles were composed of fission products, mainly radioactive cesium (Cs) as well as radioactive iron (Fe), zinc (Zn) and fuel uranium (U) [8, 9]. Iron was used in FNPP and Zn was added to the primary cooling water for regulation to reduce cobalt-60 (^{60}Co). These observations indicate that FNPP particles were formed in the nuclear reactors. FNPP particles were also found in the terrestrial environment, and their composition was similar to the particles found in the aerosol filters. FNPP particles consist primarily of highly oxidized silicate (Si) glass (a particle contains approximately 70 wt. % of SiO_2). The compositions of FNPP particles are markedly different from those in the debris [10] and radioactive Cs was condensed in almost all representative particles [11]. Details of physicochemical characteristics or classification are described in Chapter 15 of this book by Ninomiya. FNPP particles have been reported under different names and are referred to as radioactive Cs-bearing particles in this article. Radioactive Cs-bearing particles are classified into type A and B. Type A particles are typically spherical and a few micrometer in size. Type B particles are larger in size, ranging tens to hundreds of microns and are irregular but not fibrous in shape.

16.2   Movement of Insoluble Particles in the Respiratory Tract

Since the respiratory tract is in contact with the air through the breathing, radioactive Cs-bearing particles in the air could be inhaled. Therefore, understanding the motion of particles in the respiratory tract is fundamental to evaluate biological effects by inhaled particles. Particle size is a defining factor in prediction of deposition site in the respiratory system. The particles with greater momentum are likely to deposit at the site where the direction of bulk airflow changes rapidly and the larger particles mostly deposit at the bend and bifurcation in upper airways. Several models exist to predict particle motion in the respiratory tract. The International Commission on Radiological Protection (ICRP) Task Group on lung dynamics published a lung model for estimating deposition in and clearance from the respiratory tract and revised it as publication 66 in 1994 [12]. The National Council on Radiation Protection and Measurements (NCRP) independently proposed a respiratory tract dosimetry model using different empirical deposition equations from the ICRP model [13]. Yeh et al. simulated particle motion in naso-oro-pharyngolaryngeal (NOPL), tracheobronchial and pulmonary regions using both ICRP and NCRP models and compared the deposition fraction to the particle size [14]. Both models predict a similar distribution of deposition fraction in tracheobronchial and pulmonary regions for particles of 1–10 μm in size, which are type A particles. Both models also predict a similar distribution of particles between 1 and 2.5 μm in NOPL. However, the ICRP model anticipates that the deposition fraction of 2.5–10 μm particles in the NOPL region is higher than the deposition fraction predicted by the NCRP model.

Based on simulation using the ICRP model, type A particles are capable of entering the deep lung. Type A particles preferentially deposit in the NOPL region, and the particles that pass through the NOPL region subsequently deposit in the tracheobronchial and the pulmonary regions. In contrast, particles of larger than 10 μm, that is, type B particles are preferentially deposited by the settling and impaction mechanism in the NOPL region and at bends in the airflow, including the first few generations of bronchi. Most type B particles do not reach the alveoli [15].

Interaction of particles with tissues of the respiratory tract may result in deleterious effects on human health. Following deposition, it is crucial to remove these particles by physiological clearance processes. Large insoluble particles deposited in the NPOL region are predominantly removed by mucociliary clearance. In this mechanism, the mucus layer traps particles and cilia remove particles from the airways by the coordinated beating. The mucus layer also covers the surface of conducting airways to terminal bronchioles, and the insoluble particles deposited in these regions are also eliminated by mucociliary clearance. The velocity of cilia transport slows down as the particles move further into airways, and mucus transport rate is considered to be 100–600 μm/min in the terminal bronchiole and 5–20 mm/min in the trachea.

Mucociliary clearance does not occur in the gas exchange region. In this region insoluble particles are phagocytosed by alveolar macrophages and are eliminated by the mucociliary escalator [16]. Macrophage clearance depends on particle size and shape. Alveolar macrophages are less efficient at phagocytosing ultrafine particles smaller than 0.1 μm in size or fibrous particles [17–20]. Ultrafine particles retained in the gas exchange region may be translocated to the lung epithelial lining by endocytosis and by caveolae vesicles [21]. It is currently unknown if irregular shape characteristic of type B particles interferes with phagocytosis by alveolar macrophage.

Radioactive Cs-bearing particles primarily consist of SiO_2 at approximately 70 wt.% and are amorphous [22]. High occupational exposure to Si particles leads to silicosis, and internalization of amorphous Si particles into macrophages has been studied. Size-specific cellular uptake of amorphous Si particles was evaluated using macrophages experimentally differentiated from THP-1, a human acute monocytic leukemia cell line. Macrophages incorporated ultrafine particles more effectively than micronparticles. Class A scavenging receptors such as SR-A1 and SR-A6 are involved in cellular uptake of amorphous Si particles, and both receptors exhibit size-specific cellular uptake of particles with approximately 50 and 100 nm, but not 1 μm in diameter, indicating that macrophages recognize radioactive Cs-bearing particles via SR-As-independent mechanisms [23, 24]. Recently, SR-B1 was identified as a Si receptor and recognized various sizes of amorphous Si, including the particles with 1 μm in size [25]. However, SR-B1 recognizes and tethers the particles on the cell surface, but internalization into rodent lymphoma cells or fibroblasts expressing mSR-B1 did not occur. The fate of macrophage incorporating or tethering radioactive Cs-bearing particles is unknown and should be examined further.

16.3   Perspective on Biological Impact of Exposure to Radioactive Cs-Bearing Particles

The biological impact of exposure to radioactive Cs-bearing particles have become a prominent concern since discovery of these particles. Controversy relating to the biological impact of insoluble radioactive particles began in the late 1960s in connection with lung exposure to hot particles. Hot particles are physically small but cause relatively high local dose of radiation and ICRP discussed carcinogenesis in the lung exposed to hot particles.

16.3.1   Fate and the Effect of Type B Particles

Our preliminary experiments demonstrated that type B particles have a potential to affect surrounding cells. Human retinal pigmented epithelial cells immortalized by telomerase transfection were seeded at lower density (upper images in Fig. 16.1), and cell proliferation was monitored by live-cell imaging up to 53 h from the beginning of co-culture with a type B particle with more than 1000 Bq of ^{137}Cs. This 53-h period of co-culture is sufficient for the cell culture to become confluent without radioactive Cs-bearing particles. Cell density in the region not directly contacting the particle increased by the end of live-cell imaging, while the region in direct contact with the particle was sparse (bottom images in Fig. 16.1). The shape of the cells near the particle was larger and flatter than proliferating cells. These morphological changes are representative of cellular senescence, as persistent cell cycle arrest is one of the features of senescent cells. Since ionizing radiation induces cellular senescence, senescence-like phenotype may be elicited in the cells directly

(A)   (B)

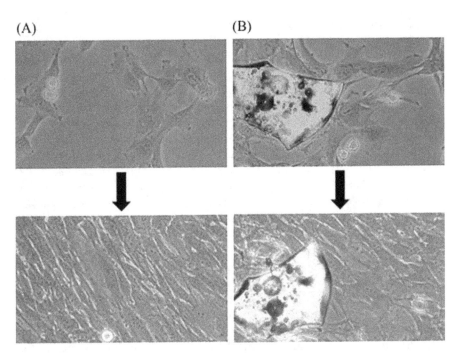

Fig. 16.1 Live-cell imaging of normal human epithelial cells
Cells not in contact with the particle (A) and those in contiguity with the particle (B) were monitored by a live-cell imaging. The images were taken at the beginning (upper) and the end (lower) of the culture. A type B particle is seen at the left side of images (B)

attached to the radioactive Cs-bearing particle. We extracted total RNA from three groups of cells; co-cultured with a type B particle, co-cultured with a nonradioactive mock particle (mock particle) and without a particle, respectively, and gene expression level for 21,088 genes was investigated by microarray analysis (3D-Gene Human Oligo chip 25k, TORAY, Kanagawa, Japan). As a control study, a mock particle composed of SiO_2 and stable cesium was generated. Expression of 1,944 genes was at least 1.5 fold different between cells co-cultured with a mock particle and those without, indicating that the physical presence of the particle influences gene expression. In total, 2,455 genes showed at least 1.5 fold different between cells treated with a radioactive Cs-bearing particle and the mock control. Type B particle increased the expression of cell cycle suppressors such as CDKN1C, CDKN1B and GADD45A, and reduced the expression of cell cycle promoters such as cyclin E2, cyclin B1 and cyclin A2. These results demonstrated that type B particles suppress cell growth, consistent with our observation using live-cell imaging. Ionizing radiation triggers activation of cell cycle checkpoints, resulting in cell cycle arrest in the G1 and G2/M phases. However, cell cycle suppressor genes upregulated by radioactive Cs-bearing particle are not involved in checkpoint regulation. Therefore, further studies are needed to elucidate how radioactive Cs-bearing particles suppress cell growth.

16.3.2 The Effect of Type A Particles

Due to the larger size of type B particles, inhaled type B particles are predicted to be removed by mucociliary clearance without entering into the deep lung region. In contrast, type A particles could enter the alveolar region. Macrophages function to recognize, take up and clear Si particles, but macrophages may fail to remove insoluble Si particles from the lung. Internalization of insoluble particles into macrophages triggers lysosomal stress, resulting in cell death pathway [26, 27]. In addition, macrophages tether Si particles on their cellular surface, but do not internalize when SR-B1 mediates recognition of Si particles [25]. The fate of the macrophage tethering insoluble particles is not clear, but the signaling pathway associated with lysosomal stress is activated in particle-bound macrophages. Regardless of whether macrophages take up or tether radioactive Cs-bearing particles, these particles may ultimately be retained in the alveolar region, resulting in the increased risk of lung injury. Furthermore, radioactive Cs-bearing particles can become a source of chronic radiation exposure to surrounding cells. Since alveolar macrophages are known to be radioresistant [28], other cells present in alveoli such as fibroblasts, epithelial cells and endothelial cells may be the most susceptible to chronic radiation exposure. When analyzing biological effects of radioactive Cs-bearing particles, bystander effects should be considered.

Radioactive Cs-bearing particles emit β-particles and γ-rays, and β-particles dissipate within a short distance. Therefore, spatial distributions of radiation doses around these particles are nonuniform and bystander effect may not be ignored. Several reports suggest that radioactive Cs-bearing particles may physically trigger cellular reactions, which may result in additive or synergistic effects for low-dose (LD) radiation and inflammatory responses [29–31]. Radiation-induced bystander effect is characterized by the appearance of radiation effects in non-irradiated cells adjacent to exposed cells. Bystander effect is mediated by either soluble factors secreted from irradiated cells [32–34] or gap junction communication which allows molecular signals to pass through protein channels between adjacent cells [35, 36]. Bystander factors are associated with various effects in adjacent non-irradiated cells such as induction of DNA damages, chromosomal aberrations, gene mutations and cell death, and alterations in epigenetic status including gene expression and cell cycle regulation [37–39]. Although a unified mechanism has not been discovered, it is assumed that molecules relating to DNA damage response such as ATM and p53 are involved in the expression of bystander effect [36, 40, 41]. Bystander effect occurs both in vivo and in vitro [42, 43]. Aside from the canonical bystander effect, a pathway from non-irradiated bystander cells to the primarily irradiated cells has been reported. For example, DNA damage in irradiated cells is mitigated by co-culture with non-irradiated bystander cells [44]. High-dose single radiation exposure induces cellular senescence in somatic cells, which is referred to stress-induced premature senescence (SIPS) [45]. LD fractionated irradiation (5 cGy) or chronic low-dose-rate (LDR) (4.1 mGy/h) radiation exposure also elicits SIPS [46, 47]. Cells undergoing SIPS secrete proteins related to inflammatory responses or radiation resistance enhancement [48, 49]. Since the SIPS phenotype is irreversible, prolonged inflammatory microenvironment induced by radioactive Cs-bearing particles has potential to contribute to lung fibrosis.

16.4   Conclusion

This article reviewed potentially hazardous biological effects of radioactive Cs-bearing particles produced by the FNPP accident. Our preliminary in vitro data demonstrated that a type B particle interrupts the growth of adjacent cells. Additive or synergistic biological effects of Si component should be considered when studying the health effect of persistent LDR radiation by the particles. Animal experiments using radioactive Cs-bearing particles are necessary to investigate the distribution of ingested particles and their tissue effects, including cancer induction.

Acknowledgment This work was supported by the grant provided by the Nuclear Energy Science & Technology and Human Resource Development Project (through Concentrating Wisdom) #281302 to MS and JSPS KAKENHI Grant #26253022 to MF from the Ministry of Education, Culture, Sports, Science and Technology of Japan.

References

1. Crocker GR, O'Connor JD, Freiling EC (1966) Physical and radiochemical properties of fall-out particles. Health Phys 12(8):1099–1104
2. Schmidt-Burbach GM (1970) Experimental determination of the distribution of dose rate around punctiform radioactive particles. Health Phys 18(3):295–296
3. Smith JN, Ellis KM, Aarkrog A et al (1994) Sediment mixing and burial of the ^{239,240}Pu pulse from the 1968 Thule, Greenland nuclear weapons accident. J Environ Radioact 25:135–159
4. Cutshall N, Osterberg C (1964) Radioactive particle in sediment from the Columbia River. Science 144(3618):536–537
5. Salbu B, Bjørnstad HE, Svarenb HE et al (1993) Size distribution of radionuclides in nuclear fuel reprocessing liquids after mixing with seawater. Sci Total Environ 130/131:51–63
6. Sandalls FJ, Segal MG, Victorova N (1993) Hot particles from Chernobyl: a review. J Environ Radioact 18:5–22
7. Osuch S, Dabrowska M, Jaracz P et al (1989) Isotopic composition of high-activity particles released in the Chernobyl accident. Health Phys 57(5):707–716
8. Adachi K, Kajino M, Zaizen Y et al (2013) Emission of spherical cesium-bearing particles from an early stage of the Fukushima nuclear accident. Sci Rep 3:2554
9. Abe Y, Iizawa Y, Terada Y et al (2014) Detection of uranium and chemical state analysis of individual radioactive microparticles emitted from the Fukushima nuclear accident using multiple synchrotron radiation X-ray analyses. Anal Chem 86(17):8521–8852
10. Burns PC, Ewing RC, Navrotsky A (2012) Nuclear fuel in a reactor accident. Science 335(6073):1184–1188
11. Satou Y, Sueki K, Sasa K et al (2016) First successful isolation of radioactive particles from soil near the Fukushima Daiichi Nuclear Power Plant. Anthropocene 14:71–77
12. International Commission on Radiological Protection (ICRP) (1994) Human respiratory tract model for radiological protection. Ann ICRP 24(1–3). ICRP publication 66

13. National Council of Radiation Protection and Measurements (NCRP) (1997) Deposition, retention, and dosimetry of inhaled radioactive substances. NCRP report no. 125
14. Yeh H-C, Cuddihy RG, Phalen RF, I-Y C (1996) Comparisons of calculated respiratory tract deposition of particles based on the proposed NCRP model and the new ICRP66 model. Aerosol Sci Technol 25:134–140
15. Zhang Z, Kleinstreuer C, Donohue JF et al (2005) Comparison of micro- and nano-size particle depositions in a human upper airway model. J Aerosol Sci 36(2):211–233
16. Lay JC, Bennett WD, Kim CS et al (1998) Retention and intracellular distribution of instilled iron oxide particles in human alveolar macrophages. Am J Respir Cell Mol Biol 18:687–695
17. Oberdörster G (1993) Lung dosimetry: pulmonary clearance of inhaled particles. Aerosol Sci Technol 18(3):279–289
18. Kreyling WG, Semmler-Behnke M, Moller W (2006) Ultrafine particle-lung interactions: does size matter? J Aerosol Med 19(1):74–83
19. Karakoti AS, Hench LL, Seal S (2006) The potential toxicity of nanomaterials—the role of surfaces. JOM 58(7):77–82
20. Stanton MF, Layard M, Tegeris A et al (1977) Carcinogenicity of fibrous glass: pleural response in the rat in relation to fiber dimension. J Natl Cancer Inst 58(3):587–603
21. Oberdörster G, Oberdörster E, Oberdörster J (2005) Nanotoxicology: an emerging discipline evolving from studies of ultrafine particles. Environ Health Perspect 113(7):823–839
22. Satou Y, Sueki K, Sasa K et al (2018) Analysis of two forms of radioactive particles emitted during the early stages of the Fukushima Daiichi Nuclear Power Station accident. Geochem J 52(2):137–143
23. Hamilton RF Jr, Thakur SA, Mayfair JK et al (2006) MARCO mediates silica uptake and toxicity in alveolar macrophages from C57BL/6 mice. J Biol Chem 281(45):34218–34226
24. Orr GA, Chrisler WB, Cassens KJ et al (2011) Cellular recognition and trafficking of amorphous silica nanoparticles by macrophage scavenger receptor A. Nanotoxicology 5(3):296–311
25. Tsugita M, Morimoto N, Tashiro M et al (2017) SR-B1 is a silica receptor that mediates canonical inflammasome activation. Cell Rep 18(5):1298–1311
26. Costantini LM, Gilberti RM, Knecht DA (2011) The phagocytosis and toxicity of amorphous silica. PLoS One 6(2):e14647
27. Thibodeau MS, Giardina C, Knecht DA et al (2004) Silica-induced apoptosis in mouse alveolar macrophages is initiated by lysosomal enzyme activity. Toxicol Sci 80(1):34–48
28. Oghiso Y, Yamada Y (1992) Heterogeneity of the radiosensitivity and origins of tissue macrophage colony-forming cells. J Radiat Res 33:334–341
29. Huh D, Matthews BD, Mammoto A et al (2010) Reconstituting organ-level lung functions on a chip. Science 328:1662–1668
30. Kusaka T, Nakayama M, Nakamura K et al (2014) Effect of silica particle size on macrophage inflammatory responses. PLoS One 9(3):e92634
31. Merget R, Bauer T, Küpper H et al (2001) Health hazards due to the inhalation of amorphous silica. Arch Toxicol 75(11–12):625–634
32. Lehnert BE, Goodwin EH (1997) Extracellular factor(s) following exposure to a particles can cause sister chromatid exchanges in normal human cells. Cancer Res 57:2164–2171
33. Mothersill C, Seymour CB (1998) Cell-cell contact during gamma irradiation is not required to induce a bystander effect in normal human keratinocytes: evidence for release during irradiation of a signal controlling survival into the medium. Radiat Res 149:256–262
34. Sowa Resat MB, Morgan WF (2004) Radiation-induced genomic instability: a role for secreted soluble factors in communicating the radiation response to non-irradiated cells. J Cell Biochem 92(5):1013–1019
35. Mancuso M, Pasquali E, Leonardi S et al (2008) Oncogenic bystander radiation effects in patched heterozygous mouse cerebellum. Proc Natl Acad Sci U S A 105(34):12445–12450
36. Azzam EI (2001) Direct evidence for the participation of gap junction-mediated intercellular communication in the transmission of damage signals from alpha-particle irradiated to nonirradiated cells. Proc Natl Acad Sci U S A 98(2):473–478

37. Nagasawa H, Little JB (2002) Bystander effect for chromosomal aberrations induced in wild-type and repair deficient CHO cells by low fluences of alpha particles. Mutat Res 508:121–129
38. Persaud R, Zhou H, Baker SE et al (2005) Assessment of low linear energy transfer radiation–induced bystander mutagenesis in a three-dimensional culture model. Cancer Res 65(21):9876–9882
39. Lyng FM, Seymour CB, Mothersill C (2002) Initiation of apoptosis in cells exposed to medium from the progeny of irradiated cells: a possible mechanism for bystander-induced genomic instability? Radiat Res 157:365–370
40. Hagelstrom RT, Askin KF, Williams AJ et al (2008) DNA-PKcs and ATM influence generation of ionizing radiation-induced bystander signals. Oncogene 27(53):6761–6769
41. Azzam EI, de Toledo SM, Gooding T et al (1998) Intercellular communication is involved in the bystander regulation of gene expression in human cells exposed to very low fluences of alpha particle. Radiat Res 150:497–504
42. Hatzi VI, Laskaratou DA, Mavragani IV et al (2015) Non-targeted radiation effects in vivo: a critical glance of the future in radiobiology. Cancer Lett 356(1):34–42
43. Chai Y, Lam RK, Calaf GM et al (2013) Radiation-induced non-targeted response in vivo: role of the TGFbeta-TGFBR1-COX-2 signalling pathway. Br J Cancer 108(5):1106–1112
44. He M, Dong C, Xie Y et al (2014) Reciprocal bystander effect between alpha-irradiated macrophage and hepatocyte is mediated by cAMP through a membrane signaling pathway. Mutat Res 763–764:1–9
45. Suzuki M, Boothman DA (2008) Stress-induced premature senescence (SIPS). J Radiat Res 49(2):105–112
46. Tsai KKC, Stuart J, Y-YE C et al (2009) Low-dose radiation-induced senescent stromal fibroblasts render nearby breast cancer cells radioresistant. Radiat Res 172(3):306–313
47. Rombouts C, Aerts A, Quintens R et al (2014) Transcriptomic profiling suggests a role for IGFBP5 in premature senescence of endothelial cells after chronic low dose rate irradiation. Int J Radiat Biol 90(7):560–574
48. Aggarwal BB, Gehlot P (2009) Inflammation and cancer: how friendly is the relationship for cancer patients? Curr Opin Pharmacol 9(4):351–369
49. Klokov D, Leskov K, Araki S et al (2013) Low dose IR-induced IGF-1-sCLU expression: a p53-repressed expression cascade that interferes with TGFbeta1 signaling to confer a pro-survival bystander effect. Oncogene 32(4):479–490

17

Immortalization of Radiation-Exposed Tissues

Tomokazu Fukuda

Abstract The evaluation of the potential biological effect of radiation is important for human health. We previously reported the deposit of radionuclides in animals from the ex-evacuation zone of the FNPP accident. In case of internal exposure, the dose of the radiation is largely affected by the metabolism of the radionuclides. We can assume that the radiation-exposed tissue is the mixed population of cells, which have genetic mutations in multiple and different sites of the genome. To detect the mutations of each cell, we need a new technology which allows us to detect the genetic alternation of the genome at the single cell level. We previously found that the expression of human-derived mutant CDK4, and overexpression of Cyclin D1 and TERT allow us to immortalize cells derived from multiple species. By applying this immortalization method to radiation-exposed tissues, we can obtain the multiple immortalized cell lines from the primary tissues. Since each cell line is expected to be derived from a single cell of the tissue, each cell line is expected to keep mutation spectrum of the genome occurring in the cell of origin after irradiation.

Keywords Fukushima Daiichi Nuclear Power Plant accident · Low-dose radiation · Mutation · Cell immortalization · Cyclin dependent kinase 4 · Cyclin D · Telomere reverse transcriptase

17.1 Background

The evaluation of the potential biological effect of radiation is important for human health. Especially, in Japan, the Fukushima Daiichi Nuclear Power Plant (FNPP) accident leads public to recognize the importance of risk assessment and radiation

T. Fukuda (✉)
Graduate School of Science and Engineering, Iwate University, Morioka, Japan

Soft-Path Engineering Research Center (SPERC), Iwate University, Morioka, Japan

safety [1]. We previously reported the deposit of radionuclides in animals from the ex-evacuation zone of the FNPP accident [2–9]. In general, radiation exposure can be classified into the two aspects, internal and external exposure. Although radiation effects on human have been analyzed by epidemiological studies on health of the atomic bomb survivors of Hiroshima and Nagasaki (Hibakusha), such as solid cancer incidence [10], these studies are mainly based on the result of acute and external exposure. However, in case of chronic and internal exposure, radiation dose is largely affected by the metabolism of the radionuclides.

The cellular effect caused by radiation exposure is represented mainly by DNA damage, such as strand breaks. Although the damaged DNA can be repaired through several biological pathways, such as homologous recombination and/or nonhomologous end joining [11], the damage of DNA caused by radiation exposure has a risk to induce the alteration of sequence information of genome, including base substitutions and nucleotide deletion. Exposure to 1 Gy of X-ray radiation reportedly induces 1,000 single strand breaks in an irradiated cell [12]. Furthermore, the clustered damaged sites on the genome can lead multiple nucleotide substitutions in the genome [12, 13]. As far as we know, the position of nucleotide substitutions or insertion/deletion mutations is random, and the sensitive area is not reported yet [14]. Therefore, in in vivo tissues, we can assume that the radiation-exposed tissue is the mixed population of cells, which have genetic mutations in multiple and different sites of the genome. To detect mutations of each cell, we need a new technology which allows us to detect genetic alternations at the single cell level.

17.2 Immortalization of Wild Macaque-Derived Cell with the Expression of Mutant Cyclin-Dependent Kinase and Cyclin D and Telomerase

Primary cells cannot proliferate infinitely due to cellular stress and senescence during the cell culture [15]. However, the expression of oncogenic proteins such as SV40 large T or E6/E7 proteins of human papilloma virus (HPV) allows us to grow a cell, which is close to immortalization [16, 17]. Although the immortalization by these oncogenic molecules is quite reproducible and efficient, the genomic and chromosomal status becomes instable and sometimes causes abnormalities. Especially, the expression of E6/E7 causes polyploid abnormality of the genome [18]. Furthermore, these methods induce the inactivation of p53 protein which is one of the most important molecules to keep the integrity of the genome and is even called as the guardian of the genome. Furthermore, the combination of shRNA of p16 and c-Myc oncogene was reported to induce the immortalization of human mammary epithelial cells (HMEC) [19]. However, even in this method, additional chemical treatment of benzo(a)pyrene to HMEC was required for the immortalization, which possibly explained by the additional genetic alteration is required for the infinite cell proliferation [19].

We previously found that the combination of expression of R24C mutant type of cyclin-dependent kinase 4 (*CDK4*), *Cyclin D1* and enzymatic complex of telomerase (*TERT*) allows us to bypass the negative feedback of the senescence protein, p16 [20]. To be noted, the amino acid sequence of the cell cycle regulators, such as CDK4 and Cyclin D are quite conservative among species. Based on this evolu-

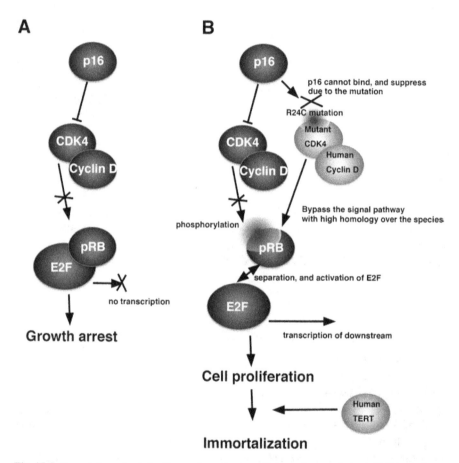

Fig. 17.1 Expected accelerated cell growth mechanism of mutant human-derived CDK4, Cyclin D and TERT over the multiple species. (**a**) Cell growth arrest under the cellular senescence and/or cellular stress. The protein level of p16 increases under the senescence. The p16 protein binds to the pocket of the CDK4 and negatively regulates the activity of CDK4-Cyclin D complex. The inactivated CDK4-Cyclin D complex cannot induce the phosphorylation of pRB resulting in its inactivation. Under the intact condition of pRB, E2F is not released from the binding status, resulting in no transcription of the downstream genes and growth arrest of the cells. (**b**) Enhanced cellular proliferation with mutant *CDK4* and *Cyclin D1* and *TERT* overexpression. Due to the R24C mutation of the human-derived CDK4, p16 protein cannot suppress the activity of protein complex of mutant CDK4 and Cyclin D. The exogenously introduced human-derived mutant CDK4-Cyclin D complex with endogenous pRB phosphorylates pRB. Due to the phosphorylation and inactivation of pRB, the transcription factor, E2F would be released from the complex and induce cell proliferation. This figure was reproduced from our previous publication with slight modification [22]

tional conservancy of the molecules, we found that the expression of human-derived mutant *CDK4* and overexpression of *Cyclin D1* and *TERT* allow us to immortalize cells derived from multiple species [21–23]. Furthermore, the expression of mutant type *CDK4* and *Cyclin D1* allows us to bypass the negative regulation of p16 and pRB (retinoblastoma protein) while keeping the function of p53 protein intact. Since p53 is an important molecule to maintain the genome, we confirmed that the chromosome condition of immortalized cells is intact in comparison with wild-type cells [24]. This situation led us to assume that the expression of mutant *CDK4*, *Cyclin D1* and *TERT* could immortalize a cell from the irradiated tissue. From the characters of introducing genes, we named the immortalized cells as the K4DT (mutant CDK4, Cyclin D1 and TERT) cells, and this immortalization technique is called as the K4DT method (Fig. 17.1).

Applying the K4DT method to radiation-exposed tissues, we can obtain multiple immortalized cell lines from the primary tissues. Since each cell line is expected to be derived from a single cell of the tissue, each cell line is expected to keep mutation spectrum of the genome occurring in the cell of origin after irradiation. The immortalized cells can be also obtained from the tissues of non-irradiated control animals.

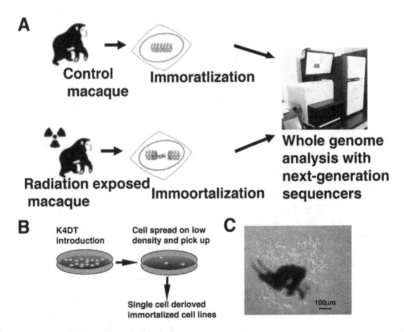

Fig. 17.2 Strategy of whole genome analysis to identify the radiation-induced genomic alteration with the next-generation sequencing. (**a**) Note that the irradiated tissue is the mixed population of cells with mutations at the random position of the genome. After the establishment of immortalized cells from a single cell, we can identify the genomic alteration caused by radiation exposure. (**b**) Strategy of a single cell-derived immortalized cells from the K4DT method. (**c**) Cell morphology which showed the proliferation as a single cell-derived colony. The colony has been marked by black marker from the bottom of the plastic dish

Based on the whole genome sequencing of cloned control and a single cell from radiation-exposed tissues, genomic mutations caused by radiation exposure could be detected. The genetic alterations at the single cell level caused by irradiation can be detected by the combination of the K4DT immortalization method and whole genome sequencing (Fig. 17.2a).

Currently, we are underway to establish immortalized cell lines derived from wild macaques with internal and external exposure to radioactive cesium from the affected area of the FNPP accident to elucidate genetic alternation caused by low-dose (LD) and low-dose-rate (LDR) radiation exposure (Fig. 17.2b). We have obtained multiple skeletal muscle-derived cell lines as part of the effort. We are interested in skeletal muscle because it is nondividing and accumulates the highest concentration of radioactive cesium in the body.

References

1. Bowyer TW, Biegalski SR, Cooper M et al (2011) Elevated radioxenon detected remotely following the Fukushima nuclear accident. J Environ Radioact 102:681–687
2. Fukuda T (2018) Estimation of concentration of radionuclides in skeletal muscle from blood, which based on the data from abandoned animals in Fukushima. Anim Sci J. https://doi.org/10.1111/asj.13018
3. Fukuda T, Hiji M, Kino Y et al (2016) Software development for estimating the concentration of radioactive cesium in the skeletal muscles of cattle from blood samples. Anim Sci J 87:842–847
4. Fukuda T, Kino Y, Abe Y et al (2015) Cesium radioactivity in peripheral blood is linearly correlated to that in skeletal muscle: analyses of cattle within the evacuation zone of the Fukushima Daiichi Nuclear Power Plant. Anim Sci J 86:120–124
5. Fukuda T, Kino Y, Abe Y et al (2013) Distribution of artificial radionuclides in abandoned cattle in the evacuation zone of the Fukushima Daiichi Nuclear Power Plant. PLoS One 8:e54312
6. Koarai K, Kino Y, Takahashi A et al (2016) 90Sr in teeth of cattle abandoned in evacuation zone: record of pollution from the Fukushima-Daiichi Nuclear Power Plant accident. Sci Rep 6:24077
7. Takahashi S, Inoue K, Suzuki M et al (2015) A comprehensive dose evaluation project concerning animals affected by the Fukushima Daiichi Nuclear Power Plant accident: its set-up and progress. J Radiat Res 56:i36–i41
8. Urushihara Y, Kawasumi K, Endo S et al (2016) Analysis of plasma protein concentrations and enzyme activities in cattle within the ex-evacuation zone of the Fukushima Daiichi Nuclear Plant Accident. PLoS One. https://doi.org/10.1371/journal.pone.0155069
9. Yamashiro H, Abe Y, Fukuda T et al (2013) Effects of radioactive caesium on bull testes after the Fukushima nuclear plant accident. Sci Rep 3:2850
10. Ozasa K, Shimizu Y, Suyama A et al (2012) Studies of the mortality of atomic bomb survivors, report 14, 1950-2003: an overview of cancer and noncancer diseases. Radiat Res 177:229–243
11. Popp HD, Brendel S, Hofmann W-K et al (2017) Immunofluorescence microscopy of γH2AX and 53BP1 for analyzing the formation and repair of DNA double-strand breaks. J Vis Exp. https://doi.org/10.3791/56617
12. Eccles LJ, O'Neill P, Lomax ME (2011) Delayed repair of radiation induced clustered DNA damage: friend or foe? Mutat Res Fundam Mol Mech Mutagen 711:134–141
13. Georgakilas AG, O'Neill P, Stewart RD (2013) Induction and repair of clustered DNA lesions: what do we know so far? Radiat Res 180:100–109

14. Adewoye AB, Lindsay SJ, Dubrova YE et al (2015) The genome-wide effects of ionizing radiation on mutation induction in the mammalian germline. Nat Commun 6:6684
15. Hayflick L, Moorhead PS (1961) The serial cultivation of human diploid strains. Exp Cell Res 25:585–621
16. Zhang H, Jin Y, Chen X et al (2007) Papillomavirus type 16 E6/E7 and human telomerase reverse transcriptase in esophageal cell immortalization and early transformation. Cancer Lett 245:184–194
17. Fukuda T, Katayama M, Yoshizawa T et al (2012) Efficient establishment of pig embryonic fibroblast cell lines with conditional expression of the simian vacuolating virus 40 large T fragment. Biosci Biotechnol Biochem 76:1372–1377
18. Bester AC, Roniger M, Oren YS et al (2011) Nucleotide deficiency promotes genomic instability in early stages of cancer development. Cell 145:435–446
19. Garbe JC, Vrba L, Sputova K et al (2014) Immortalization of normal human mammary epithelial cells in two steps by direct targeting of senescence barriers does not require gross genomic alterations. Cell Cycle 13:3423–3435
20. Shiomi K, Kiyono T, Okamura K et al (2011) CDK4 and cyclin D1 allow human myogenic cells to recapture growth property without compromising differentiation potential. Gene Ther 18:857–866
21. Fukuda T, Iino Y, Eitsuka T et al (2016) Cellular conservation of endangered midget buffalo (Lowland Anoa, *Bubalus quarlesi*) by establishment of primary cultured cell, and its immortalization with expression of cell cycle regulators. Cytotechnology 68:1937–1947
22. Fukuda T, Eitsuka T, Donai K et al (2018) Expression of human mutant cyclin dependent kinase 4, Cyclin D and telomerase extends the life span but does not immortalize fibroblasts derived from loggerhead sea turtle (*Caretta caretta*). 8:9229
23. Kuroda K, Kiyono T, Isogai E et al (2015) Immortalization of fetal bovine colon epithelial cells by expression of human cyclin D1, mutant cyclin dependent kinase 4, and telomerase reverse transcriptase: an in vitro model for bacterial infection. PLoS One 10:e0143473
24. Fukuda T, Iino Y, Eitsuka T et al (2016) Cellular conservation of endangered midget buffalo (Lowland Anoa, *Bubalus quarlesi*) by establishment of primary cultured cell, and its immortalization with expression of cell cycle regulators. Cytotechnology 68:1937–1947

18

Water with Radioactive Cs: Impacts on Mice

The Effect of Low-Dose Internal Exposure of Multigeneration Chronic Oral Intake of ^{137}Cs in Mice Offspring

Hiroo Nakajima

Abstract To investigate the transgenerational effect of chronic low-dose-rate internal radiation exposure after the Fukushima Nuclear Power Plant accident in Japan, every generation of mice was maintained in a radioisotope facility with free access to drinking water containing ^{137}CsCl (100 Bq/ml). Descendent mice were assessed for γ-H2AX foci in hepatocytes, the micronucleus test and chromosome aberration analysis in bone marrow cells, DNA mutations in the liver, tumorigenicity in the lung, oxidative stress in blood plasma, and metabolome analysis in the heart.

In this chapter, the author tries to introduce that animal experiments are useful to understand low-dose and low-dose-rate radiation effects.

Keywords Radioactive cesium · Drinking water · Mice · Transgenerational effects · Internal exposure · Low-dose radiation

18.1 Background

18.1.1 Why Have Low-Dose Radiation Issues Not Been Resolved, Despite the Passage of More Than 90 Years?

A study report by Muller in 1927 using *Drosophila* illustrated the concerns of the effect of low-dose (LD) radiation in humans on the next generation [1, 2]. Then, in the 1950s these concerns became more pressing with the expected increase in environmental radiation (global fallout) caused by nuclear tests at Bikini Atoll and the

H. Nakajima (✉)
Institute for Radiation Sciences, Osaka University, Toyonaka, Osaka, Japan
e-mail: nakajima@irs.osaka-u.ac.jp

rise of pronuclear power advocates [3]. And it had been proposed to establish United Nations Scientific Committee on the effects of Atomic Radiation (UNSCEAR) at the UN General Assembly in 1955.

Recessive lethal mutations generated by genetic damage caused by radiation-induced DNA double-strand breaks create latent lethal actions in offspring. In particular, recessive lethal mutations on the X chromosome exert a more lethal effect in males, who have only one X chromosome, than in females, who have two X chromosomes. For this reason, the male population was expected to decrease, presenting an urgent problem for humankind, and there was heated discussion on the establishment of safe ranges for public exposure [4].

There is an increased incidence of cancer in the atomic bomb survivors of Hiroshima and Nagasaki (Hibakusha) [5–8]. Gene mutations are considered as the main cause of carcinogenesis [9]. These further increased the concern over radiation-induced mutagenesis [10–13] and carcinogenesis [14]. Fortunately, there was no reduction in the male population due the effect of radiation even among Hibakusha [15, 16], but concerns over the genetic and the carcinogenic effect of low-dose (LD) radiation exposure on the next generation has remained the central proposition of the discussion [17–23].

However, even in 2018, more than 90 years since these concerns were first raised, there are no firm conclusions on the safe range of LD radiation. The scientific opinion on the effect of LD radiation has been divided extremely as seen in the debate on the effect of LD radiation on the prevalence of childhood thyroid cancer attributed to the Fukushima Daiichi Nuclear Power Plant (FNPP) accident [24, 25]. Dispute of so-called scientists on this topic in front of the general public created blunders resulting in a massive loss of faith in scientists by the general public. Unfortunately, some people misunderstand the dose limit of radiation management guidelines as the dangerous border. Furthermore, the baseless testimony by some scientists adversely affects the consensus formation between the administration and the public, which could be the field of trans-science [26]. The job of scientists is not to decide whether LD radiation is safe or dangerous but to provide evidence-based quantitative data for risk assessment [27] and publicize the data to the public in a fair and easy-to-understand manner.

Why have these issues not been resolved, despite the passage of more than 70 years? However, most of the scientists have virtually focused their research on the use of radiation as a tool, but not mainly implemented on the presence or absence of a safety range of LD radiation. There are a number of reasons for this question such as:

1. Large-scale samples and long-term observations are required to obtain data on the effect of LD radiation.
2. Physical dose measurement is rather easy, but the evaluation of absorbed dose is difficult.

3. The impact of the results could be low in spite of the laborious experiments under limited conditions, which makes scientists have little incentive for this research.

4. Difficulty to quantify confounding factors and to evaluate exposure dose in epidemiological studies.

5. Difficult to verify negative data and, as a result, to acquire research funding constantly.

It is also envisioned that there are a large volume of buried data on the effect of LD radiation that have not been disclosed to the world among research, which is because positive data are publishable but negative ones are not even well-planned and precise as the research is (particularly with animal experiments).

To resolve these issues, it is essential to motivate scientists to acquire data as a legacy for future generations, develop methods able to quantitatively evaluate radiation with a high level of sensitivity even in areas of LD radiation, and accumulate basic data that will form evidence for risk evaluation.

18.1.2   Investigating the Effect of Multigenerational Low-Dose (LD) Internal Exposure on Offspring

Concern of LD radiation exposure as a social problem immediately after the Chernobyl nuclear power plant (CNPP) accident (1986) put emphasis on the carcinogenicity of the radiation-exposed generation and the genetic impact on the next generation. Previous research on the effect of radiation exposure have been reported through the life span study (LSS) and adult health study (AHS) of Hibakusha, industrial radiation exposure [28–33], medical radiation exposure [34–36], high background exposure [37–39] and various animal experiments [40–43]. Recently studies have attempted to establish highly sensitive quantitative detection of biological reactants after exposure to LD radiation [44–46]. However, almost all of these studies focused on carcinogenesis and mutation of the first or the second generation only. It can be said that there has been no research on analyzing the effect of LD and LD-rate (LDR) internal exposure on descendants of multiple generations. Minor variations of each generation need to be accumulated over consecutive generations until the effect of radiation becomes evident, which would take more than several hundred years in humans. Therefore, the author attempted to evaluate the transgenerational effect of LD radiation within a short timeframe using inbred mice with more rapid intergenerational changes than humans. The author believes that a variety of compelling results on biological effects of LD radiation have been obtained from animal experiments in the author's laboratory. In this chapter, the author would like to introduce those results in the hope of helping readers consider the effect of persistent internal LD radiation.

18.2   Experimental Methods and Results

18.2.1   Effects of Radioactive Cesium-Containing Water on Mice

Two littermates were selected from a litter of A/J mouse strain. One group was reared with free drinking of radioactive cesium chloride (^{137}CsCl) containing water (100 Bq/ml) as the LD internal exposure group (^{137}Cs group), and the other group was reared with ^{137}Cs free intake of water as control without radiation. The offspring of these mice produced more than 15 generational changes through sibling mating (equates to approximately 300 years of generational changes in humans), and the following 1–8 experiments were carried out by selected mice from the ^{137}Cs and the control groups with the same ancestor (origin). Since manuscripts of experiments 3 to 8 are in preparation, detailed data of these experiments are not shown.

1. *Organ distribution of* ^{137}*Cs in Mice*
   Figure 18.1 shows the distribution of ^{137}Cs in the organs of small wild animals collected in a moderately contaminated region of Belarus (Babchin Village) 11 years after the CNPP accident [47, 48].
   Figure 18.2 shows the difference of ^{137}Cs concentration in organs of A/J mice that continuously drank ^{137}CsCl water (100 Bq/ml) as drinking water for 8 months in the laboratory [49]. As shown in Figs. 18.1 and 18.2, there was a large accumulation of ^{137}Cs in muscles. Also, despite drinking a constant daily amount of ^{137}Cs, the amount of ^{137}Cs in the organs was constant, reaching a plateau corresponding to the consumption amount and flattening out thereafter.

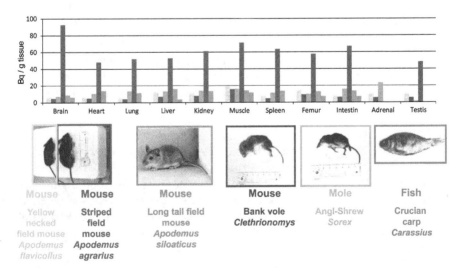

Fig. 18.1 Distribution of ^{137}Cs in the various organs of wild animals living in the middle-level contaminated areas (Babchin Village) of Belarus in 1997

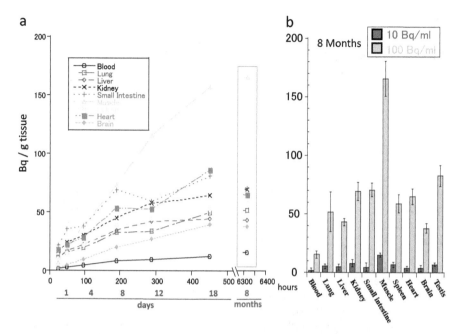

Fig. 18.2 (**a**) The accumulation of ^{137}Cs over time in each organ in male mice that had ingested water supplemented with ^{137}CsCl (100 Bq/ml). The estimated intake was 16 Bq/g body weight/day and 440 Bq/mouse/day. (**b**) Body distribution of ^{137}Cs after 8 months of ad libitum ^{137}Cs water (10 Bq/ml or 100 Bq/ml) (error bar represents 95% CI)

Figure 18.3 shows the overview of the experiment on the transgenerational effect by internal exposure to ^{137}Cs water. Figure 18.4 shows the chronic internal exposure and ^{137}Cs concentration in the liver at all stages of mouse development in this experiment.

Figure 18.5 presents the 18th-generation (8 months old) mice and shows that the amount of ^{137}Cs in all the organs rapidly attenuated after ^{137}Cs was ceased from drinking. ^{137}Cs concentration in muscles declined more slowly than that in other organs [49].

2. *Effect on Intergenerational Litter Size and Sex Ratio*

Mean sex ratio (Fig. 18.6a) and mean litter size (Fig. 18.6b) were compared between the ^{137}Cs group and the control group over generations 1–18 [49]. If the amount of intake was converted to drinking water in humans, 100 Bq/ml ^{137}Cs water for mice would equate to 100,000 Bq/L for humans. There was no effect on the mouse sex ratio and litter size even after 18 continuous generations of mice continued to drink 100 Bq/ml ^{137}Cs water [49]. In this experiment, it will be the proof that there was no decrease in the male population due to radiation effects of the fallout of nuclear tests.

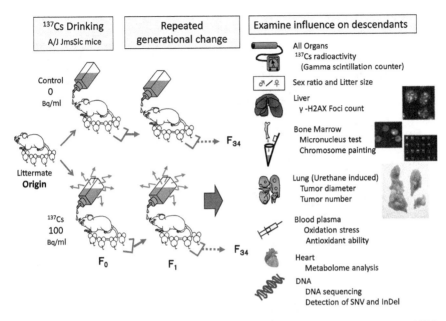

Fig. 18.3 The overview of experiments on transgenerational effects by internal exposure of ^{137}Cs

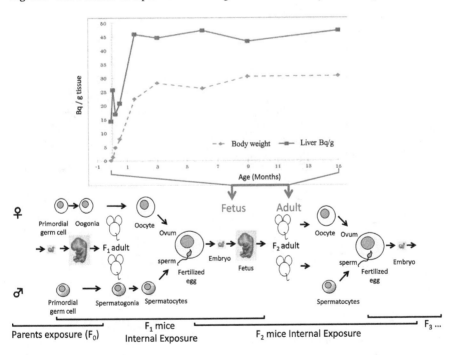

Fig. 18.4 Chronic internal exposure and ^{137}Cs concentration at all stages of mouse development. The graph shows the change of internal ^{137}Cs concentration in the liver of mouse from fetus to 15 months age

Fig. 18.5 The decrease of the ^{137}Cs level in 8-month-old female mice in the 18th generation after the mice had started drinking nonradioactive water

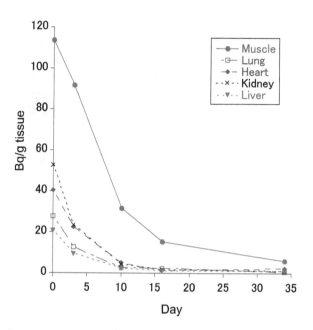

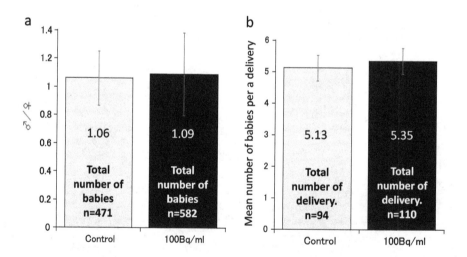

Fig. 18.6 (**a**) The mean sex ratio of all generations (F_1–F_{18}), (**b**) the mean litter size of all generations (F_1–F_{18}) (error bar represents 95% CI)

3. *Induction of DNA Double-Strand Breaks in Hepatocytes*

Measurement of the number of γ-H2AX foci, which is the indicator of the first process for repair of DNA double-strand breaks generated in cells [50–54], was performed with hepatocytes. There were significantly more DNA double-strand breaks per generation (100 days) with approximately 50 mGy (2313 Bq per mouse, mean 93.54 Bq/g body weight, 59.5 Bq per liver) of ^{137}Cs internal exposure compared to the control group.

4. *Chromosomal Abnormalities and the Micronucleus Test in Bone Marrow Cells*

Chromosomal abnormalities common to all cells were detected using the multicolor Fluorescence In Situ Hybridization (FISH) technique, and the micronucleus test was conducted in the tenth-generation mice. No common chromosomal abnormalities were found in any of bone marrow cells in the ^{137}Cs group. There was increased incidence of DNA double-strand breaks, as described in 3 (above), but common chromosomal abnormalities carried over to the next generation were not found after the tenth generation of mice. Similarly, the micronucleus test revealed no significant increase in the ^{137}Cs group.

5. *Effect of Low-Dose (LD) Radiation on Lung Carcinogenesis*

The incidence of lung tumors and the mean tumor mass in 10-month-old mice were assessed. There was no difference in the incidence of spontaneous onset and urethane-induced lung tumors between the ^{137}Cs group and the control group. Interestingly, the growth rate of tumors was significantly inhibited in the ^{137}Cs group.

6. *Oxidative Stress in Blood Plasma*

The balance of oxidative stress and antioxidant capacity in the body was studied. In the ^{137}Cs group, there was a significant elevation of plasma 8-oxodihydroguanine, which is an indicator of oxidative stress, but there was no significant difference from the control group in terms of antioxidant capacity. The generated oxidative stress was considered to be within an allowable biological range.

7. *Metabolome Analysis in the Heart*

The glycolytic pathway in the ^{137}Cs group was inhibited, and there was also a reduction in antioxidants such as glutathione (GSH) and cysteine.

8. *Quantitative Analysis of DNA Base Mutations by Whole Genome Sequencing in the Liver*

Detection of base sequence mutations in noncoding regions, where base mutations generated in germ cells do not affect life support but are expected to accumulate in each subsequent generation, was attempted in male mice of F_{20} and F_{23} origin. There was a higher incidence of single-nucleotide variant (SNV) and insertion/deletion (Indel) per 3 billion bases at intron and intergenic sites than at exon sites by comparing whole genome sequence in each generation. However, even with repeated generations, there was not a large difference in the total base mutation rate between the ^{137}Cs group and the control group.

18.2.2 Detecting Carcinogenesis and Mutations Caused by Oxidative Stress

The fatal effect of radiation is the DNA double-strand break. In the high-dose range of high linear energy transfer (LET) radiation, dominant process of the breaks is caused by direct action, while in low LET radiation such as X-rays, about two-thirds of the biologic damages are caused by indirect action through radicals generated by reaction with water in the cell [55]. In case of LD and LDR radiation, the action exerting the greatest effect is assumed to be oxidative stress caused by reactive oxygen species attributed to radiation exposure. Reports indicate that using genetically modified mice (DNA mismatch repair-deficient mice) that are unable to repair the DNA damage enables highly sensitive detection of DNA damage caused by oxidative stress agents and radiation [56, 57]. Also, these mice show a high incidence of small intestine tumors through oral consumption of an oxidative stress agent potassium bromide (KBr) aqueous solution [58, 59]. The author has started quantitative research on carcinogenesis and mutations caused by oxidative stress using this mouse strain.

18.2.3 Comparison of Internal Exposure and External Exposure

The effect of exposure in humans is based on the absorbed radiation dose in Gy (J/kg). However, this unit is obtained from the energy absorbed per kg of tissue. Internal exposure experiments using mice are different from external exposure ones, so the absorbed radiation dose from the radioactive substance must be considered in terms of the amount accumulated in a body that weighs less than 1 kg (around 0.025 kg). However, most of the γ-rays from a mouse body measuring approximately 3 cm in diameter completely pass out of the body without transferring the energy. In these instances, is it a good idea to evaluate with Gy, which is an absorption dose to the order of kg? This same question can be applied to humans. It is vital to confirm if the effect of internal exposure can be evaluated with a dose conversion factor from Bq to Gy in same way in infants and adults where there are differences in the distribution and size of organs. The dose conversion factor from Bq to Gy for rats is shown by ICRP (pub108), but no information for mice. If the actual internal exposure dose is not accurately evaluated in the mouse experiment, it significantly affects the quality of outcome of the experiment. To determine the dose conversion factor of the mice in these experimental conditions, it can be considered three evaluation methods. Previous evaluation methods for the internal exposure dose include the dose estimated from physical calculations (3.14 μGy.g/Bq.day) and evaluation methods using Monte Carlo analysis such as the EGS5 code system (Electron

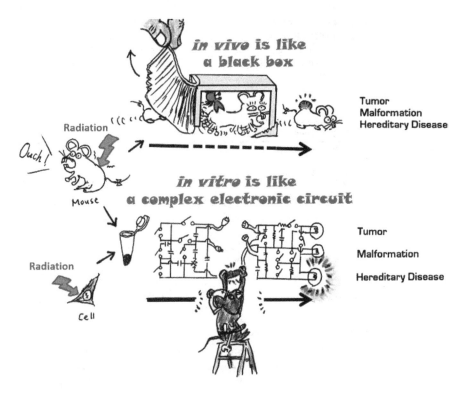

Fig. 18.7 The strategy gap in vivo and in vitro

Gamma Shower Version 5) (3.00 μGy.g/Bq.day, evaluated by Endo D, Rakuno Gakuen Univ.) and the PHITS (Particle and Heavy Ion Transport Code System) (3.40 μGy.g/Bq.day: evaluated by Endo S, Hiroshima Univ.). We are continuing research to establish methods for evaluating internal exposure dose with an attempt to adopt new techniques such as quantifying noncoding RNA (ncRNA) specific to the reaction against radiation exposure [45]. If it is possible to detect a dose-dependent reaction with highly sensitive and precise quantification of the biological reactions caused by radiation, then it would be a great help to verify equivalent dose of internal and external exposure.

There are strategy gap between in vivo and in vitro experiments (Fig.18.7). "In vivo" is like a black box, and "in vitro" is like a complex electronic circuit. We want to know what is going on inside the black box, and we want to know the pathway of circuit from the beginning to the end. Particularly in studies of LD radiation effects, bridging the gap between in vivo and in vitro experiments can effectively lead to clear results.

Table 18.1 Homology between mouse and human. Spontaneous mutation rate per generation, base substitution rate per nucleotide per generation and the number of genes in mice are almost the same as humans. This is the reason that the mice can be used as the experimental animal to know hereditary influence in humans. [b]Average of base substitution rate in references of human [58–68]

	Mutation rate per generation	Base substitution rate per nucleotide per generation	Generation interval (months)	Number of genes	Homology with the human DNA sequence
Human	3.6×10^{-6a}	1.2×10^{-8b}	300	30,000	
Mouse	6.1×10^{-6a}	0.54×10^{-8c}	3	30,000	97%
Drosophila (fruit flies)	1.8×10^{-6a}		0.3	14,000	60%

[a][58]
[b][59–68]
[c][69]

18.3 Conclusion

The experiments with *Drosophila* made people aware of the effect of radiation on human genetics. The mouse is a closer experimental animal to humans. However, the data obtained from inbred mouse strains are the same as those obtained from monozygotic twins, as all the mice have the same genetic background, so the obtained data reflect genetic bias of one person, irrespective of the number of mice used. Therefore, inbred mice may not be suitable for regulatory science research for the general public, where a variety of exposure disorders are envisioned. However, mice have almost the same spontaneous mutation rate per generation, base substitution rate per nucleotide per generation, and number of genes as humans (Table 18.1) [60–71]. Therefore, the inbred mouse experiment is effective for removing confounding factors and for detecting fundamental biological effects, which would be inconceivable in humans.

Obtaining data that contribute to the establishment of the radiation safety range has not necessarily been vigorously implemented for more than 70 years since the nuclear test at the Bikini Atoll. Those data will also be extremely effective to determine if an unpredictable radiation exposure occurs. We hope that many scholars will undertake research on the effect of LD radiation throughout the world.

Acknowledgment This work was partly supported by the Japan Society for the Promotion of Science (JSPS) [KAKENHI Grant Numbers JP23310037, JP26253022, JP26550039] and Research (project) on the Health Effects of Radiation organized by Ministry of the Environment, Japan.

References

1. Muller HJ (1927) Artificial transmutation of the gene. Science 66:84–87
2. Muller HJ (1928) The measurement of gene mutation rate in Drosophila, its high variability, and its dependence upon temperature. Genetics 13:279
3. Federation of American Scientist, "Proposal for a United Nations Commission to study the problem of H-bomb test," February 16, 1955, in folder: "Studies on effect of 1955," Box 232, Records relating to atomic energy matters, 1948–1962, special assistant for the secretary of state for atomic energy and outer space, Records of the department of state, National Archives at College Park, USA
4. Schull WJ, Neel JV (1958) Radiation and the sex ratio in man. Science 128(3320):343–348
5. Oho G (1956) New statistical observation of malignant neoplastic death in A-bomb survivors. Jpn Med J 1686:8–18. in Japanese
6. Pierce DA, Shimizu Y, Preston DL et al (1996) Studies of the mortality of atomic bomb survivors. Report 12, Part I. Cancer: 1950–1990. Radiat Res 146(1):1–27
7. Preston DL, Ron E, Tokuoka S et al (2007) Solid cancer incidence in atomic bomb survivors: 1958–1998. Radiat Res 168(1):1–64
8. Richardson D, Sugiyama H, Nishi N et al (2009) Ionizing radiation and leukemia mortality among Japanese atomic bomb survivors, 1950–2000. Radiat Res 172(3):368–382
9. Deman J, Van Larebeke N (2001) Carcinogenesis: mutations and mutagens. Tumour Biol 22(3):191–202
10. United Nations Scientific Committee on the effects of atomic radiation. Radiation-induced chromosome aberrations in human cells. UNSCEAR (1969) Reports 1969. United Nations, New York, pp 98–155
11. Sasaki MS, Miyata H (1968) Biological dosimetry in atomic bomb survivors. Nature 220:1189–1193
12. Russel WL (1955) Genetic effects of radiation in mice and their bearing on the estimation of human hazards. Proc Int Conf Peacef Uses Atom Energy 11:382–383. United Nation, 1956
13. Sankaranarayanan K (1982) Genetic effects of ionizing radiation in multicellular eukaryotes and the assessment of genetic radiation hazards in man. Elesevier, Amsterdam
14. Little JB (2000) Radiation carcinogenesis. Carcinogenesis 21(3):397–404
15. Schull WJ, Neel JV, Hashizume A (1966) Some further observations on the sex ratio among infants born to survivors of the atomic bombings of Hiroshima and Nagasaki. Am J Hum Genet 18(4):328–338
16. Neel JV, Schull WJ (1991) Children of atomic bomb survivors. A genetic study. National Academy Press, Washington, DC
17. Yoshimoto Y, Neel JV, Schull WJ et al (1990) The frequency of malignant tumors during the first two decades of life in the offspring of atomic bomb survivors, RERF technical repot, pp 4–90 (Am J Hum Genet, 1990; 46:1041–1052)
18. Furitsu K, Ryo H, Yeliseeva KG et al (2005) Microsatellite mutations show no increases in the children of the Chernobyl liquidators. Mutat Res 581:69–82
19. Fujiwara S, Suyama A, Cologne JB et al (2008) Prevalence of adult-onset multifactorial disease among offspring of atomic bomb survivors. Radiat Res 170:451–457
20. Kodaira M, Roy H, Kamata N et al (2010) No evidence of increased mutation rates at microsatellite loci in offspring of A-bomb survivors. Radiat Res 173:205–213
21. Tatsukawa Y, Cologen JB, Hsu WL et al (2013) Radiation risk of indvidual multifactorial diseases in offspring of the atomic-bomb survivors: a clinical health study. J Radiol Prot 33:281–293
22. Izumi S, Koyama K, Soda M et al (2003) Cancer incidence in children and young adults did not increase relative to parental exposure to atomic bomb. Br J Cancer 89:1709–1713
23. Sankaranarayanan K, Chakraborty R (2000) Ionizing radiation and genetic risks XI. The doubling dose estimates from the mid-1950s to the present and the conceptual change to the use

of human data on spontaneous mutation rates and mouse data on induced mutation rates for doubling dose calculations. Mutat Res 453:107–127

24. Tsuda T, Tokinobu A, Yamamoto E et al (2016) Thyroid cancer detection by ultrasound among residents ages 18 years and younger in Fukushima, Japan: 2011 to 2014. Epidemiology 27:316–322

25. Ohira T, Takahashi H, Yasumura S et al (2016) Comparison of childhood thyroid cancer prevalence among 3 areas based on external radiation dose after the Fukushima Daiichi nuclear power plant accident. The Fukushima health management survey. Medicine 95(35):e4472. https://doi.org/10.1097/MD.0000000000004472

26. Weinberg AM (1972) Science and trans-science. Minerva 10(2):209–222

27. Sankaranarayanan K, Nikjoo H (2015) Genome-based, mechanism-driven computational modeling of risks of ionizing radiation: the next frontier in genetic risk estimation? Mutat Res 764:1–15

28. Glenn JA Jr, Galindo J, Lawrence CE (1960) Chronic radium poisoning in a dial painter. Case report. Am J Roentgenol Radium Therapy, Nucl Med 83:465–473

29. Woodard HQ, Higinbotham NL (1962) Development of osteogenic sarcoma in a radium dial painter thirty-seven years after the end of exposure. Am J Med 32:96–102

30. Gardner MJ, Snee MP, Hall AJ et al (1990) Results of case-control study of leukaemia and lymphoma among young people near Sellafield nuclear plant in West Cumbria. BMJ 300:423–429

31. Workfold R (2002) Cancer in offspring after paternal preconceptional irradiation. J Radiol Prot 22:191–194

32. Johnson KJ, Alexander BH, Doody MM et al (2008) Childhood cancer in the offspring born in 1921–1984 to US radiologic technologists. Br J Cancer 99(3):545–550

33. Leuraud K, Richardson DB, Cardis E et al (2015) Ionising radiation and risk of death from leukaemia and lymphoma in radiation-monitored workers (INWORKS): an international cohort study. Lancet Haematol 2(7):e276–e281

34. Robison LL, Armstrong GT, Boice JD et al (2009) The childhood cancer survivor study: a National Cancer Institute-supported resource for outcome and intervention research. J Clin Oncol 27(14):2308–2318

35. Madanat-Harjuoja LM, Malila N, Lähteenmäki PM et al (2010) Preterm delivery among female survivors of childhood, adolescent and young adulthood cancer. Int J Cancer 127(7):1669–1679

36. Pearce MS, Salotti JA, Little MP et al (2012) Radiation exposure from CT scans in childhood and subsequent risk of leukaemia and brain tumours: a retrospective cohort study. Lancet 380(9840):499–505

37. Jiang T, Hayata I, Wang C et al (2000) Dose-effect relationship of dicentric and ring chromosomes in lymphocytes of individuals living in the high background radiation areas in China. J Radiat Res 41(Suppl):63–68

38. Tao Z, Zha Y, Akiba S et al (2000) Cancer mortality in the high background radiation areas of Yangjiang, China during the period between 1979 and 1995. J Radiat Res 41(Suppl):31–41

39. Zhang W, Wang C, Chen D et al (2003) Imperceptible effect of radiation based on stable type chromosome aberrations accumulated in the lymphocytes of residents in the high background radiation area in China. J Radiat Res 44(1):69–74

40. Ishii K, Hosoi Y, Yamada S et al (1996) Decreased incidence of thymic lymphoma in AKR mice as a result of chronic, fractionate low dose total-body X irradiation. Radiat Res 146:582–585

41. Courtade M, Billote C, Gasset G et al (2002) Life span, cancer and non-cancer disease in mouse exposed to a continuous very low dose of γ-irradiation. Int J Radiat Biol 78:845–855

42. Tanaka S, Tanaka IB, Sasagawa S et al (2003) No lengthening of life span in mice continuously exposed to gamma rays at very low dose rates. Radiat Res 160(3):376–379

43. Tang FR, Loke WK, Khoo BC (2017) Low-dose or low-dose-rate ionizing radiation–induced bioeffects in animal models. J Radiat Res 58(2):165–182

44. Nakajima H, Ohno M, Ishihara H (2018) Quantitative assessment for the effects of chronic low-dose internal Cesium-137 radiation exposure on genomic, carcinogenic and hereditary effects

in mice. Report for Study (Group) of the Health Effects of Radiation Organized by Ministry of the Environment, Japan, p 143. https://www.env.go.jp/chemi/rhm/reports/h2903e_3.pdf

45. Ishihara H, Tanaka I, Yakumaru H et al (2016) Quantification of damage due to low-dose radiation exposure in mice: construction and application of a biodosimetric model using mRNA indicators in circulating white blood cells. J Radiat Res 57(1):25–34

46. Manens L, Grison S, Bertho JM et al (2016) Chronic exposure of adult, postnatal and in utero rat models to low-dose [137]Cesium: impact on circulating biomarkers. J Radiat Res 57(6):607–619

47. Nakajima H, Ryo H, Yamaguchi Y et al (2000) Biological concentration of radionuclides in plants and animals after the Chernobyl catastrophe. In: Sato F, Yamada Y, Onodera J (eds) Biological effects of low dose radiation. Institute for Environmental Sciences, Aomori, pp 199–205

48. Nakajima H, Saito T, Ryo H et al (2008) Ecological decrease and biological concentration of radionuclides in plants and animals after the Chernobyl catastrophe. In: Miura T, Kinoshita N (eds) Proceedings of the eighth workshop on environmental radioactivity. High Energy Accelerator Research Organization (KEK), Proceedings 2007–2016, Tsukuba, pp 113–118

49. Nakajima H, Yamaguchi Y, Yoshimura Y et al (2015) Fukushima simulation experiment: assessing the effects of chronic low-dose internal ^{137}Cs radiation exposure on litter size, sex ratio, and biokinetics in mice. J Radiat Res 56:i29–i35. Special Issue-Fukushima

50. Rogakou EP, Pilch DR, Orr A et al (1998) DNA double-stranded breaks induce histone H2AX phosphorylation on serine 139. J Biol Chem 273:5858–5868

51. Rogakou EP, Boon C, Redon C et al (1999) Megabase chromatin domains involved in DNA double-strand breaks *in vitro*. J Cell Biol 146:905–915

52. Nakajima H, Tsuboi R, Nomura T (2002) Biodosimetry by detecting H2AX foci in peripheral WBC and SCID lymphoma cell line after ^{137}Cs gamma-rays irradiation (abstract). J Radiat Res 43:445

53. Rothkamm K, Löbrich M (2003) Evidence for a lack of DNA double-strand break repair in human cells exposed to very low x-ray doses. Proc Natl Acad Sci U S A 100:5057–5062

54. Nakajima H, Tsuboi R, Nomura T (2003) Biodosimetry by detecting H2AX foci in human peripheral lymphocytes and mouse organ tissues after [137]Cs γ-ray irradiation (abstract). J Radiat Res 44:406

55. Hall EJ, Giaccia AJ (2019) Radiobiology for the radiologist, 8th edn. Wolters Kluwer, USA, Philadelphia, p p10. (9781496335418)

56. Egashira A, Yamauchi K, Yoshiyama K et al (2002) Mutational specificity of mice defective in the MTH1 and/or the MSH2 genes. DNA Repair 1(11):881–893

57. Ohno M, Sakumi K, Fukumura R et al (2014) 8-oxoguanine causes spontaneous de novo germline mutations in mice. Sci Rep 4:4689

58. Tsuzuki T, Piao J, Isoda T et al (2011) Oxidative stress-induced tumorigenesis in the small intestine of various types of DNA repair-deficient mice. Health Phys 100(3):293–294

59. Piao J, Nakatsu Y, Ohno M et al (2014) Mismatch repair deficient mice show susceptibility to oxidative stress-induced intestinal carcinogenesis. Int J Biol Sci 10(1):73–79

60. Drost JB, Lee WR (1995) Biological basis of germline mutation: comparisons of spontaneous germline mutation rates among drosophila, mouse, and human. Environ Mol Mutagen 25(Suppl 26):48–64

61. Roach JC, Glusman G, Smit AFA et al (2010) Analysis of genetic inheritance in a family quartet by whole-genome sequencing. Science 328:636–639

62. Conrad DF, Keebler JE, DePristo MA et al (2011) Variation in genome-wide mutation rates within and between human families. Nat Genet 43:712–714

63. Campbell CD, Chong JX, Malig M et al (2012) Estimating the human mutation rate using autozygosity in a founder population. Nat Genet 44:1277–1281

64. Kong A, Frigge ML, Masson G et al (2012) Rate of de novo mutations and the importance of father's age to disease risk. Nature 488:471–475

65. Michaelson JJ, Shi Y, Gujral M et al (2012) Whole-genome sequencing in autism identifies hot spots for de novo germline mutation. Cell 151:1431–1442

66. Segurel L, Wyman MJ, Przeworski M (2014) Determinants of mutation rate variation in the human germline. Annu Rev Genomics Hum Genet 15:47–70
67. Besenbacher S, Liu S, Izarzugaza JM et al (2015) Novel variation and de novo mutation rates in population-wide de novo assembled Danish trios. Nat Commun 6:5969
68. Rahbari R, Wuster A, Lindsay SJ et al (2016) Timing, rates and spectra of human germline mutation. Nat Genet 48:126–133
69. Wong WS, Solomon BD, Bodian DL et al (2016) New observations on maternal age effect on germline de novo mutations. Nat Commun 7:10486
70. Jónsson H, Sulem P, Kehr B et al (2017) Parental influence on human germline de novo mutations in 1,548 trios from Iceland. Nature 549:519
71. Uchimura A, Higuchi M, Minakuchi Y et al (2015) Germline mutation rates and the long-term phenotypic effects of mutation accumulation in wild-type laboratory mice and mutator mice. Genome Res 25:1125–1134

Part VI
Fukushima Accident: A Review

19

Biological Impacts of FNPP Accident: A Study of Pale Grass Blue Butterfly

Joji M. Otaki

Abstract The pale grass blue butterfly (*Zizeeria maha*) has been used as an indicator for the biological effect of the Fukushima Daiichi Nuclear Power Plant (FNPP) since 2011. Various biological aspects have been examined that point out the biological impact of the FNPP accident through the morphological abnormalities and death of the butterfly. However, the mechanisms responsible for such biological effect in the contaminated field remain elusive. In this article, previous and current studies on the pale grass blue butterfly in response to the FNPP accident are concisely reviewed to provide new directions for answering questions about the field effect.

Keywords Fukushima Daiichi Nuclear Power Plant accident · Field effect · Field-laboratory paradox · Low-dose exposure · Pale grass blue butterfly

19.1 Introduction

The pale grass blue butterfly (*Zizeeria maha*) is a common butterfly found in humanized environments throughout Japan (except Hokkaido, the northernmost major island). Larvae of this butterfly basically eat only one weed species, *Oxalis corniculata*, which is common in various environments including agricultural villages and metropolitan cities. The pale grass blue butterfly overwinters as larvae, the larvae pupate in the early spring, and adult butterflies come out at the end of April or beginning of May. Its life cycle is complete in approximately a month, and the butterfly produces several generations per year.

We chose this butterfly to study color pattern development and evolution many years before the Fukushima Daiichi Nuclear Power Plant (FNPP) accident occurred

J. M. Otaki (✉)
The BCPH Unit of Molecular Physiology, Department of Chemistry, Biology and Marine Science, Faculty of Science, University of the Ryukyus, Okinawa, Japan
e-mail: otaki@sci.u-ryukyu.ac.jp

in 2011. We established a standard rearing method [1] and studied a phenotypically plastic aspect of the wing color patterns as an important field case of "real-time" evolution in response to environmental changes [2, 3]. The color pattern changes (called modifications) that are induced in response to cold shock and observed in the northern range-margin population (Fukaura, Aomori Prefecture) have been established as phenotypically plastic traits in response to environmental changes and their genetic assimilation in the population [2–5]. In contrast, color pattern changes (called aberrations or abnormalities) that are different from modifications can also be produced by introducing genetic mutations experimentally [6]. Both cases are employed to understand the mechanisms of color pattern development in this butterfly [7].

Following the FNPP accident, we decided to apply this butterfly system to probe the biological impact of the FNPP accident [8]. We recently demonstrated quantitatively that the pale grass blue butterfly is an excellent indicator for human-living environments in terms of sampling efficiency [9]. In this article, our findings are summarized, and new directions for answering emerging questions are presented. Because of a strict page limitation, we have almost exclusively focused on our own studies and are unable to refer to other important contributions to this field.

19.2   Three Research Objectives

The first objective of our research is to clarify the presence and the cause of the biological aberration that followed the FNPP accident. Because any biological aberration might have been caused by multiple factors, the causal demonstration is not straightforward. We referred to the Koch's postulate for infectious diseases and proposed the postulates of pollutant-induced biological impacts [10]. The postulates are a set of six criteria that must be met to demonstrate the effect of the pollutants in question (e.g., radionuclides and their associated substances released from FNPP): (1) spatial relationship, (2) temporal relationship, (3) direct exposure, (4) phenotypic variability and spectrum, (5) experimental reproduction (external exposure), and (6) experimental reproduction (internal exposure) [10]. To our knowledge, the pale grass blue butterfly is the sole organism that has met the postulates in the case of the FNPP accident.

The second objective of our research is to understand the mechanisms of the biological effect caused by the FNPP accident. Mechanistic research has just begun in our laboratory, but this objective will be the focus of our laboratory for years to come.

The third objective of our research is to establish a system to monitor human-living environments throughout Japan and around the world in the future using the pale grass blue butterfly and other similar butterfly species [11–14]. In Japan, following a future nuclear accident or large-scale chemical pollution event, the pale grass blue butterfly around an accident site may be collected, and their data may be readily compared with the reference data set already available [11, 12]. In Asia, Australia, and Africa, there are butterflies similar to the pale grass blue butterfly, and they can be used as indicator species [13, 14].

19.3   Field Surveys and Experiments in 2011

In this section, we briefly review our first paper on the biological effect of the FNPP accident published in 2012 [8]. We first performed a field sampling attempt in early May 2011 to catch the first adult generation within the contaminated localities (where people continued to live) after the FNPP accident. We made another field sampling attempt in September 2011 for comparison.

We first measured the forewing size of collected butterflies. Size reduction was observed in the May male samples only. The male forewing size was inversely correlated with the ground radiation dose. Morphological abnormalities were checked for all samples; the abnormality rate was not high in May but increased in September. In September, the abnormality rate was highly correlated with the ground radiation dose.

We obtained the offspring (F_1) generation from morphologically normal female individuals caught at various localities. Their eclosion time (defined as the number of days from the first egg deposition to eclosion) and abnormality rate of appendages were both inversely correlated with the distance from FNPP. We further obtained the subsequent (F_2) generation. Abnormality rate of the F_2 generation was relatively high, and similar types of abnormalities with the F_1 generation were observed, suggesting the inheritance of abnormal traits through genetic damage.

We then used the pale grass blue butterfly from Okinawa (approximately 1,700 km southwest of FNPP and located at the southernmost tip of Japan), the least contaminated locality in Japan. An external exposure experiment using a cesium-137 (^{137}Cs) radiation source resulted in the smaller forewing size and dose-dependent survival curves. We also performed an internal exposure experiment, in which contaminated leaves from several localities were fed to larvae. Again, we observed the smaller forewing size and dose-dependent survival curves.

Together, the biological effect was demonstrated in the field samples, in the reared generations, and in the experimental reproductions. When this study was published, we received a large number of comments and criticism from around the world. Responses to these comments and criticism have been compiled, together with more detailed data analyses [15]. In that paper, we also newly presented the normal specimens of the pale grass blue butterfly caught before the FNPP accident. A thought-provoking discussion on this issue is also found in a commentary paper from other experts [16].

19.4   Additional Exposure Experiments

We quantitatively analyzed the amount of leaves and radioactive Cs that a larva ate in the internal exposure experiment [17]. Furthermore, we collected contaminated leaves with relatively low-level radioactive Cs from additional localities and performed additional internal exposure experiments [18]. As expected, we obtained dose-dependent mortality rates when larvae were fed the contaminated leaves. When the contaminated leaves were given to the next-generation larvae, the survival rate was low, but when noncontaminated leaves were given, the survival rate was high, despite the fact that their parents ate contaminated leaves. However, forewing size was inversely correlated with the cumulative amount of ingested radioactive Cs in two generations (parent and offspring).

The mortality and abnormality data that were obtained in these studies were compiled together and subjected to mathematical model fits [19]. The total abnormality data (including both mortality and abnormality) fit a Weibull function well; it is a sigmoidal curve used to describe a failure of mechanical or biological machinery. The data also fit a power function well.

Similarly, further external exposure experiments were performed. We collected contaminated soils, on which larvae from Okinawa were reared to simulate the external environment in contaminated localities. Again, we observed biological effects of external exposures (unpublished data).

19.5   Field Surveys for 3 Years and the Japan-Wide Surveys

The first study published in 2012 [8] contained the result of the field surveys that were performed in the spring and fall of 2011. We continued similar surveys twice a year until 2013 [20]. Generally, the total abnormality rate peaked in the fall of 2011 and the spring of 2012 and then decreased to normal levels by the fall of 2013. This decrease cannot solely be attributed to the decrease of contaminating radioactive Cs. The decrease probably indicates purifying selection against genetic mutants caused by the initial exposures from the FNPP accident. In addition, we revealed that the decrease in the total abnormality rate was, at least partially, a consequence of adaptive evolution of the pale grass blue butterfly to a contaminated environment [21]. The results indicate the importance of prompt surveys after nuclear (and probably other) accidents.

We also performed the Japan-wide surveys to understand the geographical distribution patterns of the forewing size with the help of Japanese amateur lepidopterists [22] because the forewing size decrease detected in the spring of 2011 should be evaluated in comparison with those from other parts of Japan. The northern populations at a high latitude had relatively small forewing sizes, but the border was located at the prefectural boundary between Fukushima and Miyagi Prefectures.

We further performed the Japan-wide surveys to understand the geographical distribution patterns of the abnormality rate and the wing color pattern modification rate [11, 12]. There were no localities that had exceptionally high abnormality or modification rates except the northern populations. The northern populations had relatively high abnormality and modification rates, but the borders for these rates were located at the prefectural boundary between Miyagi and Iwate Prefectures.

In relation to the geographical borders of both the forewing size and the abnormality and modification rates, the Fukushima populations were categorized into the southern population. The fact that the smaller forewing size and the high abnormality rates were detected in the contaminated Fukushima populations after the FNPP accident cannot thus be attributed to natural phenotypic variation in the Fukushima populations.

19.6   Internal Exposure Experiment Using the Cabbage White Butterfly

The internal exposure experiments discussed above raised the question of whether this high sensitivity of the pale grass blue butterfly is simply an exceptional case. To answer this question, we used the cabbage white butterfly (*Pieris rapae*). Larvae of this butterfly eat cabbage leaves. We cultivated cabbage using contaminated soils collected from Fukushima to feed larvae from Okinawa. We observed morphological abnormalities in this butterfly [23]. Importantly, the percentage of granulocytes among hemocytes in this butterfly was inversely correlated with radioactive Cs concentration of cabbage leaves ingested but not with the potassium radioactivity concentration [23]. These results demonstrated that the high sensitivity of the pale grass blue butterfly to internal exposures was not exceptional. The results also indicated adverse cellular changes in hemocytes in response to ingested artificial radioactive Cs. The possibility of different biological consequences between radioactive potassium and radioactive Cs is also interesting.

19.7   Toxicology of Cesium

We evaluated the effect of nonradioactive Cs as a chemical substance by feeding larvae of the pale grass blue butterfly an artificial diet containing nonradioactive Cs chloride and demonstrated its low toxicity (unpublished data). Similarly, when radioactive Cs chloride (CsCl) was fed via an artificial diet, its toxicity was also low in the pale grass blue butterfly (unpublished data), in comparison with the field-collected leaves discussed above. The difference may be called the field-laboratory paradox, but it can be explained by the contributions of the biologically mediated indirect field effect (see below). To the best of our knowledge, the pale grass blue butterfly is the sole organism that has both field-based and laboratory-based experimental results in addition to the field surveys and various types of examinations in the FNPP accident.

19.8   A 2016 Survey for Heavily Contaminated Localities

In 2016, we performed a new field survey for heavily contaminated localities that were not included in our previous surveys. This survey was made possible because these localities have been made open to everybody since 2016 due to the completion of decontamination efforts; the Japanese government wants people to come back to the area. We found that the butterfly still showed the adverse effect of the accident in these localities in 2016 (unpublished data). Furthermore, the possible effect of radioactive iodine-131 (^{131}I) and ^{137}Cs for initial acute exposures was indicated statistically (unpublished data), suggesting that the major effect of initial radiation exposure was genetic damage.

19.9   Detection of Genetic Damage

Initial exposure by several radionuclides including ^{131}I and ^{137}Cs might have damaged DNA directly. As discussed above, genetic damage caused by the FNPP accident was first suggested by the rearing experiments for the F_1 and F_2 generations in 2011 [8]. Wing color pattern abnormalities in the field-caught and reared individuals who were phenotypically similar to those of genetic mutants also suggested genetic damage in the contaminated populations [6]. We also found that the adult abnormality rate was correlated well with the ground radiation dose using data points only from the contaminated localities surveyed in the fall of 2011 and the spring of 2012, suggesting that the abnormality peak at these time points are likely caused by the initial exposures that introduced genetic damage [24]. We have not published any molecular data to date, but we hope that we will be able to present them in the future.

One of the methods used to probe genome-wide DNA polymorphism is called amplified fragment-length polymorphism (AFLP) analysis. The genomic DNA was digested with restriction enzymes into numerous fragments, which were then amplified by polymerase chain reaction (PCR). Distribution patterns of the PCR fragment lengths were then analyzed by electrophoresis. We found that the abnormality rate was correlated with genetic diversity of the population and that genetic diversity was correlated with radiation dose (unpublished data). It appears that random mutations that were introduced by radiation exposure contributed to genetic diversity and then the abnormality rate.

Furthermore, the construction of the genome database for the pale grass blue butterfly, including the RNA-seq (whole transcriptome next generation sequencing) data for nontreated individuals, is under way. These data may be compared with the RNA-seq data from irradiated individuals to check for differences in gene expression patterns.

19.10   Overwintering Larvae and Dosimetric Analysis

To evaluate the contribution of the initial exposures immediately after the FNPP accident to the biological impact quantitatively, it is important to calculate absorbed doses for the pale grass blue butterfly, considering both γ- and β-rays [25]. Such a dosimetric analysis necessarily involves mathematical simulations that introduce many assumptions. To reduce the number of biological assumptions, it is important to understand the developmental stage of the pale grass blue butterfly and its surrounding environment at the time of the FNPP accident. We performed a field survey for overwintering larvae in March 2018, and we will continue this effort in 2019. Based on the larval size and environmental information, it will be possible to calculate the initial absorbed doses that larvae received, which may clarify the field-laboratory paradox to some extent.

19.11   Field Effect via Biological Indirect Pathways

The field-laboratory paradox may be solved by the field effect, which has been ignored by most radiation scientists. There may be several types of the field effect that are mediated by biological indirect pathways [24, 26–28]. For example, the host plant leaves may change their chemical components upon irradiation; toxic chemicals may be synthesized, or some important nutrients such as vitamins may not be synthesized. These chemical changes in host plants may cause death or abnormalities in larvae. Studies on the mechanisms of these pathways are now under way in the "model ecosystem" involving the pale grass blue butterfly and its host plant.

Acknowledgments The author is grateful to the editor, Prof. Fukumoto, for his invitation to this book project and for his comments on this article. The Fukushima Project at the BCPH Unit of Molecular Physiology was financially supported partially by the Incentive Project of the University of the Ryukyus, partially by the following private foundations, Takahashi Industrial and Economic Research Foundation, the Nohara Foundation, the Sumitomo Foundation, the Asahi Glass Foundation, and act beyond trust, and partially by public donors who kindly helped us to perform our study. The Fukushima Project was also supported by volunteer amateur lepidopterists and nonlepidopterists who provided invaluable samples (including butterflies and plants) and comments. The author also thanks the current and previous members of the BCPH Unit of Molecular Physiology for their contributions to the Fukushima Project.

References

1. Hiyama A, Iwata M, Otaki JM (2010) Rearing the pale grass blue *Zizeeria maha* (Lepidoptera, Lycaenidae): toward the establishment of a lycaenid model system for butterfly physiology and genetics. Entomol Sci 13:293–302
2. Otaki JM, Hiyama A, Iwata M et al (2010) Phenotypic plasticity in the range-margin population of the lycaenid butterfly *Zizeeria maha*. BMC Evol Biol 10:252

3. Hiyama A, Taira W, Otaki JM (2012) Color-pattern evolution in response to environmental stress in butterflies. Front Genet 3:15
4. Buckley J, Bridle JR, Pomiankowski A (2010) Novel variation associated with species range expansion. BMC Evol Biol 10:382
5. Gilbert SF (2014) Developmental biology, 10th edn. Sinauer Associates, Sunderland
6. Iwata M, Hiyama A, Otaki JM (2013) System-dependent regulations of colour-pattern development: a mutagenesis study of the pale grass blue butterfly. Sci Rep 3:2379
7. Iwata M, Taira W, Hiyama A et al (2015) The lycaenid central symmetry system: color pattern analysis of the pale grass blue butterfly *Zizeeria maha*. Zool Sci 32:233–239
8. Hiyama A, Nohara C, Kinjo S et al (2012) The biological impacts of the Fukushima nuclear accident on the pale grass blue butterfly. Sci Rep 2:570
9. Hiyama A, Taira W, Sakauchi K et al (2018) Sampling efficiency of the pale grass blue butterfly *Zizeeria maha* (Lepidoptera: Lycaenidae): a versatile indicator species for environmental risk assessment. J Asia Pac Entomol 21:609–615
10. Taira W, Nohara C, Hiyama A et al (2014) Fukushima's biological impacts: the case of the pale grass blue butterfly. J Hered 105:710–722
11. Hiyama A, Taira W, Iwasaki M et al (2017) Geographical distribution of morphological abnormalities and wing color pattern modifications of the pale grass blue butterfly in northeastern Japan. Entomol Sci 20:100–110
12. Hiyama A, Taira W, Iwasaki M et al (2017) Morphological abnormality rate of the pale grass blue butterfly *Zizeeria maha* (Lepidoptera: Lycaenidae) in southwestern Japan: a reference data set for environmental monitoring. J Asia Pac Entomol 20:1333–1339
13. Gurung RD, Iwata M, Hiyama A et al (2016) Comparative morphological analysis of the immature stages of the grass blue butterflies *Zizeeria* and *Zizina* (Lepidoptera: Lycaenidae). Zool Sci 33:384–400
14. Iwata M, Matsumoto-Oda A, Otaki JM (2018) Rearing the African grass blue butterfly *Zizeeria knysna*: toward the establishment of a bioindicator in African countries. Afr Study Monogr 39:69–81
15. Hiyama A, Nohara C, Taira W et al (2013) The Fukushima nuclear accident and the pale grass blue butterfly: evaluating biological effects of long-term low-dose exposures. BMC Evol Biol 13:168
16. Møller AP, Mousseau TA (2013) Low-dose radiation, scientific scrutiny, and requirements for demonstrating effects. BMC Biol 11:92
17. Nohara C, Hiyama A, Taira W et al (2014) The biological impacts of ingested radioactive materials on the pale grass blue butterfly. Sci Rep 4:4946
18. Nohara C, Taira W, Hiyama A et al (2014) Ingestion of radioactively contaminated diets for two generations in the pale grass blue butterfly. BMC Evol Biol 14:193
19. Taira W, Hiyama A, Nohara C et al (2015) Ingestional and transgenerational effects of the Fukushima nuclear accident on the pale grass blue butterfly. J Radiat Res 56:i2–i8
20. Hiyama A, Taira W, Nohara C et al (2015) Spatiotemporal abnormality dynamics of the pale grass blue butterfly: three years of monitoring (2011–2013) after the Fukushima nuclear accident. BMC Evol Biol 15:15
21. Nohara C, Hiyama A, Taira W et al (2018) Robustness and radiation resistance of the pale grass blue butterfly from radioactively contaminated areas: a possible case of adaptive evolution. J Hered 109:188–198
22. Taira W, Iwasaki M, Otaki JM (2015) Body size distributions of the pale grass blue butterfly in Japan: size rules and the status of the Fukushima population. Sci Rep 5:12351
23. Taira W, Toki M, Kakinohana K, Sakauchi K, Otaki JM (2019) Developmental and hemocytological effects of ingesting Fukushima's radiocesium on the cabbage white butterfly *Pieris rapae*. Sci Rep 9:2625
24. Otaki JM, Taira W (2018) Current status of the blue butterfly in Fukushima research. J Hered 109:178–187

25. Endo S, Tanaka K, Kajimoto T et al (2014) Estimation of β-ray dose in air and soil from Fukushima Daiichi Power Plant accident. J Radiat Res 55:476–483
26. Garnier-Laplace J, Geras'kin S, Delta-Vedova C et al (2013) Are radiosensitivity data derived from natural field conditions consistent with data from controlled exposures? A case study of Chernobyl wildlife chronically exposed to low dose rates. J Environ Radioact 121:12–21
27. Otaki JM (2016) Fukushima's lessons from the blue butterfly: a risk assessment of the human living environment in the post-Fukushima era. Integr Environ Assess Manag 12:667–672
28. Bréchignac F (2016) The need to integrate laboratory- and ecosystem-level research for assessment of the ecological impact of radiation. Integr Environ Assess Manag 12:673–676

20

Fukushima and Chernobyl: A Comparative Study

Tetsuji Imanaka

Abstract Although both the Fukushima Daiichi Nuclear Power Plant (FNPP) accident in 2011 and the Chernobyl NPP Unit 4 (CNPP) accident in 1986 are classified as Level 7, the worst nuclear incidence on the International Nuclear and Radiological Event Scale by the International Atomic Energy Agency, there are various differences between the two, including the accident process, released radionuclide composition, and meteorological and geological conditions. The amounts of iodine-131 (^{131}I) and cesium-137 (^{137}Cs) released into the atmosphere were about six times smaller after the FNPP accident compared to the CNPP accident. Cesium-137 is the most important radionuclide in considering long-term effects of nuclear accidents. According to Chernobyl laws in Ukraine, Belarus and Russia, depending on the level of ^{137}Cs contamination, the contaminated territories were classified as alienation zone (>1480 kBq m^{-2}), obligatory resettlement zone (555–1480 kBq m^{-2}) and voluntarily resettlement zone (185–555 kBq m^{-2}). The areas of the corresponding zones around FNPP were 272, 459 and 1405 km^2, respectively, which were 11–15 times smaller compared to the CNPP accident. Meanwhile, the number of affected people around FNPP was only three to five times smaller compared to CNPP, reflecting the higher population density for the FNPP accient. Cumulative exposures for the 1st year 1 m above ground (normalized to the initial ^{137}Cs deposition of 1 MBq m^{-2}) were 63 mGy for the FNPP accident, while it was 500 mGy for the CNPP accident because more various radionuclides were emitted in case of the CNPP accident than the FNPP accident. Cumulative exposures at 30 years were evaluated to be 500 mGy and 970 mGy for the FNPP accident and the CNPP accident, respectively.

Keywords Fukushima Daiichi Nuclear Power Plant accident · The Chernobyl accident · Radioactivity release · Radioactive contamination · Cesium-137

T. Imanaka (✉)
Institute for Integrated Radiation and Nuclear Science, Kyoto University, Kumatori, Sennan, Osaka, Japan
e-mail: imanaka@rri.kyoto-u.ac.jp

20.1 Introduction

Both the Fukushima Daiichi Nuclear Power Plant (FNPP) accident in 2011 and the Chernobyl NPP Unit 4 (CNPP) accident in 1986 are classified as Level 7, which is the highest level on the International Nuclear and Radiological Event Scale (INES) as defined by the International Atomic Energy Agency (IAEA) [1, 2]. In both accidents, a large amount of radionuclides were released into the environment from the damaged reactors and large areas of land were heavily contaminated to the extent that many people have to be evacuated for a long period [3–5]. Although it is clear that each accident had a big impact on society, they were different in various aspects, including the accident process, composition of radioactive contamination and geological conditions [6, 7].

FNPP was the first nuclear power plant built by the Tokyo Electric Power Company (TEPCO) and its first unit (Unit 1: 460 MWe) began operation in 1971. By 2011, FNPP had six boiling water reactor (BWR) units (Units 2–5: 780 MWe; Unit 6: 1100 MWe) developed by General Electric (GE, USA). Units 4, 5 and 6 were out of operation at the time of the earthquake (14:46 March 11, 2011) due to annual maintenance work, while Units 1, 2 and 3 were operating at full power [8, 9].

The epicenter of the Great East Japan Earthquake was approximately 180 km away from FNPP. At 14:47, the three operating reactors were automatically shut down due to a large seismic acceleration, and emergency diesel generators (EDGs) were then activated to provide necessary electricity to the station. The tsunami waves, at over 10 m high, arrived at FNPP around 15:36 and flooded the basement of the turbine buildings where EDGs were located. EDG failure resulted in power loss for the pumps providing coolant water to remove decay heat from the reactor cores, which was the real emergency that led to the FNPP accident.

Several emergency cooling systems that do not require electric power were installed at each BWR in the event of a power outage. These cooling systems included isolation condenser (IC) systems, reactor core isolation cooling (RCIC) systems and high-pressure coolant injection (HPCI) systems. Unit 1 was equipped with IC and HPCI, while Units 2 and 3 were equipped with RCIC and HPCI. These emergency cooling systems were not designed to work for a long period, and consequently three FNPP reactors operating at the time of the tsunami became damaged one by one. The sequence of reactor damage is summarized below [10]:

- Unit 1: After EDG power failure, both IC and HPCI systems lost function. Without emergency cooling, the reactor core began meltdown, and fuel melted through the reactor pressure vessel (RPV) in the evening of March 11. At 02:30 March 12, the drywell (DW) inner pressure was measured to be 840 kPa, about twice the maximum design pressure of 427 kPa. To avoid rupture of DW, the operator tried to vent the pressure, successfully releasing pressure to the acceptable level at 14:30. At 15:36 March 12, however, a hydrogen explosion occurred at the roof of the reactor building, which was strong enough to destroy the roof and the wall on the highest floor of the reactor building.

- Unit 3: After power loss, RCIC remained functional until 11:36 March 12, and then HPCI was automatically actuated. In the early morning on March 13, the Unit 3 operator decided to switch the cooling system from HPCI to the line using fire engine water prepared outside the building. However, the cooling systems were not switched quickly, which left the reactor without a cooling supply for about 7 h. The meltdown and melt-through process of Unit 3 began in the morning on March 13 and subsequently worsened. A hydrogen explosion occurred at 11:01 March 14.
- Unit 2: After power loss, RCIC remained functional until 13:25 March 14. The Unit 2 operator attempted a change of cooling system to fire engines, but was unable to switch systems quickly, leaving the reactor without a cooling supply for about 2 h. The meltdown and melt-through process of Unit 2 began in the evening on March 14. A high DW pressure was observed that night, and an attempted venting operation was unsuccessful. In the morning on March 15, a sudden drop of DW pressure was observed, which suggested containment rupture and massive release of radioactivity into the atmosphere.

In this paper, we compare the radiological consequences between the FNPP accident and the CNPP accident with respect to the amount of radioactivity released into the atmosphere and the radioactive contamination on land.

20.2   Radioactivity Release

A gradual increase in radiation levels was observed at the entrance gate of FNPP beginning 04:00 March 12. Radioactivity release from the Unit 1 reactor building began early in the morning of March 12. The first large release occurred as a result of the Unit 1 vent operation at 14:30 March 12, followed by the hydrogen explosion at 15:36. Unit 3 began to release radioactivity in the morning on March 13. Serious radioactivity release from Unit 2 occurred from the evening of March 14 to the morning of March 15. Significant radioactivity release into the atmosphere continued up to the end of March.

The amount of radionuclides released into the environment is the basic information required to consider the scale of nuclear accidents. To date, two methods have been applied to estimate the amount of radionuclides released by the FNPP accident. The first method is a computer simulation of the accident process. This method requires many assumptions about the parameters used in the simulation, which increases uncertainty in the results [11, 12]. The second method is based on an inversion technique that combines environmental measurements and a simulation of atmospheric transport of released radionuclides [13–16]. The time trend of radioactivity release of four main radionuclides xenon-133 (^{133}Xe), ^{131}I, tellurium-132 (^{132}Te) and ^{137}Cs obtained by the second method is shown in Fig. 20.1. The data for ^{133}Xe are taken from the UNSCEAR report [4], while the data for other radionuclides are taken from the recent work by Katata et al. [16]. The cumulative distribution of radionuclide release is plotted in Fig. 20.2. Xenon-133 release was completed by

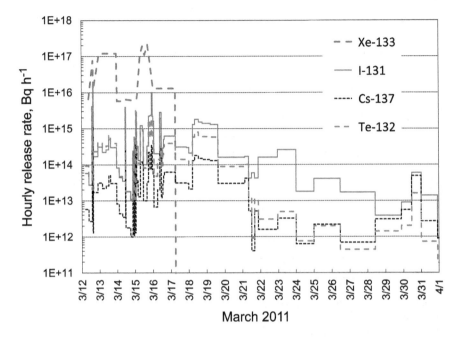

Fig. 20.1 Hourly radioactivity release from the FNPP accident into the atmosphere during March 2011

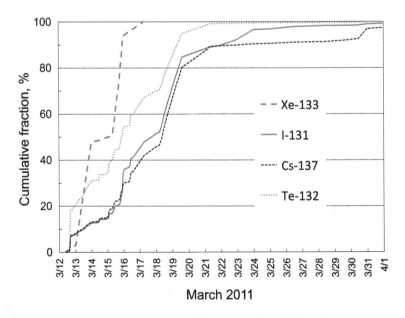

Fig. 20.2 Cumulative distribution of radionuclide release of the FNPP accident

Table 20.1 Comparison of the amounts of main radionuclides released into the atmosphere by the FNPP and CNPP accidents

Radionuclide	FNPP[a]		CNPP[b]	
	Released radioactivity, PBq	Fraction of core inventory[c], %	Released radioactivity [17], PBq	Fraction of core inventory[d], %
^{133}Xe	10,500 [4]	87	6500	100
^{131}I	300 [16]	5	1760	55
^{132}Te	310 [16]	3.6	1150	43
^{134}Cs	15 [16]	2.1	47	26
^{137}Cs	15 [16]	2.1	85	30
^{90}Sr	0.14 [12]	0.03	10	5
^{95}Zr	0.017 [12]	2×10^{-4}	84	1.5
^{103}Ru	7.5×10^{-6} [12]	9×10^{-8}	168	3.5
^{106}Ru	2.1×10^{-6} [12]	9×10^{-8}	73	3.5
^{140}Ba	3.2 [12]	0.03	240	5
^{141}Ce	0.018 [12]	2×10^{-4}	84	1.5
^{239}Np	0.076 [12]	7×10^{-5}	400	1.5
^{239}Pu	3.2×10^{-6} [12]	1×10^{-4}	0.013	1.5

[a]Decay-corrected at 14:46 on March 11, 2011
[b]Decay-corrected on April 26, 1986
[c]Inventory values from Nishihara et al. [18]
[d]Inventory values from the Ukraine report [19]

March 17. As shown in Fig. 20.2, about 80% of the release of two important radionuclides, ^{131}I and ^{137}Cs, occurred between March 15 and March 21. Radioactive materials released by two hydrogen explosions on March 12 and March 14 did not make a significant contribution to the total release.

Estimated total amounts of various radionuclides released into the atmosphere by the FNPP accident are compared with those released by the CNPP accident in Table 20.1. Xenon-133 release was greater for the FNPP accident than the CNPP accident. The released activity ratio of FNPP to CNPP for ^{131}I and ^{137}Cs is about one-sixth. Compared with the CNPP accident, very small amounts of other radionuclides such as strontium-90 (^{90}Sr), zirconium-95 (^{95}Zr), ruthenium-103 (^{103}Ru), etc. were released by the FNPP accident. These differences can be explained by the element characteristics and accident processes. Xenon-133, a rare gas radionuclide, easily escaped into the environment in both the accidents. The difference in ^{133}Xe release simply reflects the reactor power of FNPP (Units 1, 2 and 3: total 2 GW electricity) and CNPP (Unit 4: 1 GW). Because the CNPP accident was a power surge accident, the explosion occurred within the reactor core and destroyed the reactor and the building at the same time. This led to direct exposure of the damaged reactor core to the atmosphere, as well as dispersion of nuclear fuels around the damaged Unit 4 building. Meanwhile, the explosions in the FNPP accident did not happen in the reactor cores. The meltdown and melt-through of reactor cores occurred inside the containment structures without direct exposure to the atmosphere. Therefore, mainly gaseous and volatile radionuclides were released into the atmosphere in case of the FNPP accident.

20.3 Radioactive Contamination

The area of ^{137}Cs-contaminated land is the most important factor in determining the long-term effects of nuclear accidents. The first effort to make a ^{137}Cs deposition map around FNPP was carried out by a team from the US National Nuclear Security Administration (NNSA) that arrived at the Yokota air base near Tokyo in the early morning on March 16, 2011. Beginning March 17, the team conducted an aerial measuring system (AMS) survey of radioactive contamination in the area around FNPP [20]. The results of the survey were published in their website in the autumn of 2011 [21]. Dr. Sawano, an expert on the Geographic Information System (GIS) technique, found the AMS survey results for FNPP by chance and edited the ^{137}Cs deposition map as shown in Fig. 20.3 [22]. His comparison of the ^{137}Cs-contaminated area and population size for the FNPP accident and the CNPP accident is shown in Table 20.2 [23].

As seen in Table 20.2, the ^{137}Cs-contaminated area was 11–15 times larger for the CNPP accident than the FNPP accident, while the affected population living in contaminated zones defined by Chernobyl laws was only 3–5 times larger for the CNPP accident than that for the FNPP accident, reflecting a higher population density in

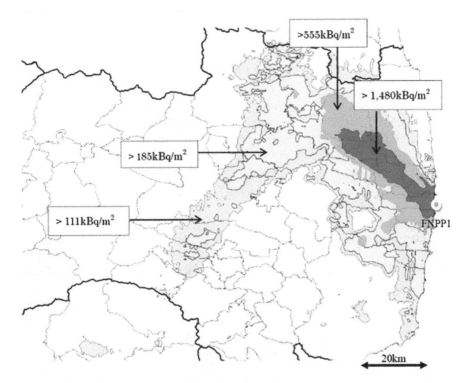

Fig. 20.3 Cesium-137 deposition map for the FNPP accident derived from DOE/NNSA AMS data following the Chernobyl classification scheme

Table 20.2 Comparison of contaminated land area and population for the FNPP and the CNPP accidents

	Zone classification in CNPP by ^{137}Cs contamination level[a]					
	(First zone)		(Second zone)		(Third zone)	
	>1480 kBq m^{-2}		555–1480 kBq m^{-2}		185–555 kBq m^{-2}	
	Area, km^2	Population, persons	Area, km^2	Population, persons	Area, km^2	Population, persons
FNPP	272	30,159	495	52,157	1405	261,076
CNPP	3100	149,000[b]	7200	235,000	19,120	689,000
CNPP/FNPP ratio	11.4	4.9	14.5	4.5	13.6	2.6

[a]According to the Chernobyl laws in Ukraine, Belarus and Russia, first zone, second zone and third zone correspond to areas of alienation, obligatory resettlement and voluntarily resettlement, respectively [23]
[b]This number includes 116,000 persons who were evacuated from the 30 km zone just after the accident in 1986

Table 20.3 Relative deposition ratios of radionuclides ([137]Cs = 1), contributing γ-ray exposure 1 m above ground [6, 7]

Radionuclide	Half-life	Exposure rate conversion factor (nGy h^{-1})/(kBq m^{-2})	Relative deposition ratio to [137]Cs	
			FNPP	CNPP
[95]Zr	65.5 days	2.82	–	20
[95]Nb	35.0 days	2.92	–	20
[103]Ru	39.3 days	1.85	–	16
[131]I	8.04 days	1.49	11.5	18
[132]Te	3.25 days	0.79	8	28
[132]I	(2.30 h)[a]	8.61	8	28
[134]Cs	2.07 years	5.97	1	0.4
[137]Cs	30.1 years	2.18	1	1
[140]Ba	12.8 days	0.57	–	22
[140]La	(1.68 days)[a]	7.83	–	11
[239]Np	2.36 days	0.60	–	120

[a]These radionuclides are treated at radioactive equilibrium with parent radionuclides

the latter. Another difference is that the east side of FNPP is surrounded by the Pacific Ocean, and the wind direction over the Japanese islands is predominantly to the east. Therefore, radionuclides were more likely to deposit in the Pacific Ocean than on land.

Gamma-ray exposure rates above the contaminated ground were calculated assuming the radionuclide composition both for the FNPP and the CNPP accidents. The deposited amounts of radionuclides contributing γ-ray exposure were shown in Table 20.3 as values of relative deposition ratio to [137]Cs [6, 7]. Deposition was assumed to occur at a time on April 26, 1986, the day of the accident for CNPP and on March 15, 2011, the day when the most severe contamination occurred for FNPP. Figure 20.4a shows the temporal change in the γ-ray exposure rate 1 m above

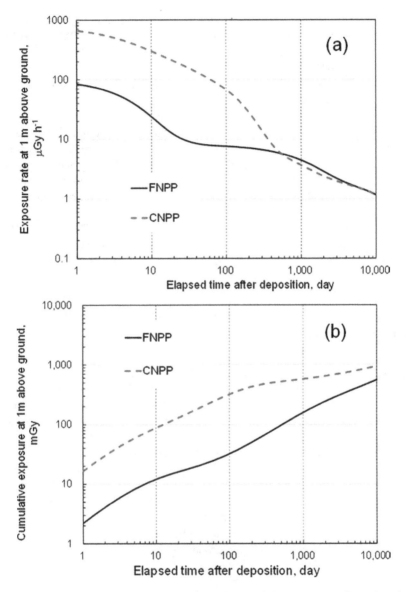

Fig. 20.4 (a) Radiation exposure rate, μGy h^{-1} and (b) cumulative exposure, mGy, at 1 m above ground normalized to the initial ^{137}Cs deposition of 1 MBq m^{-2}

ground normalized to the initial ^{137}Cs deposition of 1 MBq m^{-2} for both the FNPP and the CNPP accidents. The initial exposure rate 1 day after deposition was 80 μGy h^{-1} for the FNPP accident, while it was 700 μGy h^{-1} for the CNPP accident (nine times larger). Given the radionuclide compositions in Table 20.2, such radionuclides as ^{95}Zr, ^{103}Ru, ^{140}Ba and ^{239}Np have a substantial contribution to radiation exposure during the 1st year after the CNPP accident. Two months after deposition, radiocesiums ^{134}Cs and ^{137}Cs accounted for 99% of the radiation exposure rate in the FNPP accident, while the contribution of radiocesiums in the CNPP accident was only 4% of the exposure rate at the same time [6]. Interestingly, the FNPP exposure rate exceeded CNPP approximately 1.5 years after deposition and was almost the same after 10 years (Fig. 20.4a). This exposure rate trend is created by differences in the deposition ratio of ^{134}Cs (half-life: 2.07 years) to ^{137}Cs (half-life: 30.1 years) for the FNPP accident (^{134}Cs:^{137}Cs = 1:1) and the CNPP accident (^{134}Cs:^{137}Cs = 0.5:1).

Cumulative exposures 1 m above ground normalized to the initial ^{137}Cs deposition of 1 MBq m^{-2} were plotted both for the FNPP and the CNPP accidents (Fig. 20.4b). Cumulative exposures after the 1st year were 63 mGy and 500 mGy for the FNPP and the CNPP accident, respectively. Cumulative exposures at 30 years are 500 mGy and 970 mGy for the FNPP and the CNPP accident, respectively. The contribution of radiocesiums to the cumulative exposure for the 1st year was 83% for FNPP and 7.4% for CNPP, and at 30 years it is 98% for FNPP and 84% for CNPP. Therefore, different compositions of ground contamination between the FNPP accident and the CNPP accident led to different patterns of radiation exposure, primarily during the 1st year.

20.4   Conclusion

Although both the FNPP accident and the CNPP accident are classified as Level 7, the worst nuclear incidence classification of the INES by IAEA, there are various differences between the accidents. These differences include the accident process, released radionuclide composition and meteorological and geological conditions. Cesium-137 is the most important radionuclide to consider due to its long-term effects. The amount of ^{137}Cs released by the FNPP accident is estimated to be about six times smaller compared to the CNPP accident. The ^{137}Cs-contaminated land area is 11–15 times smaller for the FNPP accident compared to the CNPP accident. Cumulative radiation exposure for 1 year above ground at the same level of ^{137}Cs contamination is about eight times smaller for the FNPP accident compared to the CNPP accident, while it decreases to about two times for 30 years, reflecting the different radionuclide composition of ground deposition.

References

1. IAEA (2013) INES: the international nuclear and radiological event scale. User's Manual 2008 Edition. IAEA. https://www-pub.iaea.org/MTCD/Publications/PDF/INES2013web.pdf. Accessed 15 Apr 2018
2. METI (2011) METI news release. April 12, 2011. http://warp.ndl.go.jp/info:ndljp/pid/3514506/ www.nisa.meti.go.jp/english/files/en20110412-4.pdf. Accessed 15 Apr 2018
3. NAIIC (2012) The official report of The Fukushima Nuclear Accident Independent Investigation Commission. http://warp.da.ndl.go.jp/info:ndljp/pid/3856371/naiic.go.jp/en/report/. Accessed 15 Apr 2018

4. UNSCEAR (2014) Levels and effects of radiation exposure due to the nuclear accident after the 2011 great east-Japan earthquake and tsunami. Annex A of UNSCEAR 2013 Report. http://www.unscear.org/docs/reports/2013/13-85418_Report_2013_Annex_A.pdf. Accessed 15 Apr 2018
5. IAEA (2015) The Fukushima Daiichi accident. Report by the Director General. IAEA. 2015. https://www-pub.iaea.org/books/IAEABooks/10962/The-Fukushima-Daiichi-Accident. Accessed 15 Apr 2018
6. Imanaka T, Hayashi G, Endo S (2015) Comparison of the accident process, radioactivity release and ground contamination between Chernobyl and Fukushima-1. J Radiat Res 56(suppl):i56–i61. https://doi.org/10.1093/jrr/rrv074. Accessed 15 Apr 2018
7. Imanaka T (2017) Fukushima-1 and Chernobyl: comparison of radioactivity release and contamination. In: Korogodina VL et al (eds) Genetics, evolution and radiation. Hidelberg. Springer, p 225–236
8. Nuclear Emergency Response Headquarter of Japanese Government (2011) Report of Japanese Government to the IAEA ministerial conference on nuclear safety – the accident at TEPCO's Fukushima Nuclear Power Stations. http://www.japan.kantei.go.jp/kan/topics/201106/iaea_houkokusho_e.html. Accessed 15 Apr 2018
9. TEPCO (2011) Fukushima nuclear accident analysis report: interim report. December 2, 2011. http://www.tepco.co.jp/en/press/corp-com/release/betu11_e/images/111202e14.pdf. Accessed 15 Apr 2018
10. Atomic Energy Society of Japan (2014) The Fukushima Daiichi nuclear accident: final report of the AESJ Investigation Committee. Hidelberg. Springer
11. JNES (2011) A tentative report on the progression and status of the accident at the Fukushima Dai-ichi Nuclear Power Station. Tokyo. Japan Nuclear Energy Safety Organization. (in Japanese)
12. NISA (2011) Report on the situation of reactor core of Unit-1, -2 and 3 of FDNPP. (in Japanese) http://dl.ndl.go.jp/view/download/digidepo_6017222_po_20110606-1nisa.pdf?contentNo=1. Accessed 15 Apr 2018
13. Chino M, Nakayama H, Nagai H et al (2011) Preliminary estimation of release amounts of 131I and 137Cs accidentally discharged from Fukushima Daiichi Nuclear Power Plant into the atmosphere. J Nucl Sci Technol 48:1129–1134
14. Stohl A, Siebert P, Wotawa G et al (2012) Xenon-133 and caesium-137 releases into the atmosphere from the Fukushima Dai-ichi nuclear power plant: determination of the source term, atmospheric dispersion, and deposition. Atmos Chem Phys 12:2313–2343. https://doi.org/10.5194/acp-12-2313-2012
15. Terada H, Katata G, Chino H et al (2012) Atmospheric discharge and dispersion of radionuclides during the Fukushima Dai-ichi Nuclear Power Plant Accident. Part II: verification of the source term and analysis of regional-scale atmospheric dispersion. J Environ Radioact 112:141–154
16. Katata G, Chino M et al (2015) Detailed source term estimation of the atmospheric release for the Fukushima Daiichi Nuclear Power Station accident by coupling simulations of an atmospheric dispersion model with an improved deposition scheme and oceanic dispersion model. Atmos Chem Phys 15:1029–1070. https://doi.org/10.5194/acp-15-1029-2015
17. Chernobyl Forum (2005) Environmental consequences of the Chernobyl accident and their remediation: twenty years of experience. Report of the UN Chernobyl Forum Expert Group "Environment" (EGE), IAEA. http://www-ns.iaea.org/downloads/rw/meetings/environ-consequences-report-wm-08.05.pdf. Accessed 15 Apr 2018
18. Nishihara K, Iwamoto H, Suyama K (2012) Estimation of fuel compositions in Fukushima-Daiichi Nuclear Power Plant. JAEA-Data/Code 2012-018. https://jopss.jaea.go.jp/pdfdata/JAEA-Data-Code-2012-018.pdf. Accessed 15 Apr 2018
19. MinChernobyl (1996) Ten years after the accident at Chernobyl NPP: national report of Ukraine. (in Russian)

20. Lyons C, Colton D (2012) Aerial measuring system in Japan. Health Phys 102:509–515
21. US DOE/NNSA (2011) US DOE/NNSA response to 2011 Fukushima incident: raw aerial data and extracted ground exposure rates and cesium deposition. http://catalog.data.gov/dataset/us-doe-nnsa-response-to-2011-fukushima-incident-raw-aerial-data-and-extracted-ground-expos-20e73. Accessed 15 Apr 2018
22. Sawano N (2013) Really useful contamination map. Shueisha, Tokyo. (in Japanese)
23. Sawano N (2017) Mapping and analysis of ^{137}Cs ground contamination due to the FNPP accident using US NNSA AMS data. Kagaku 87:294–301. in Japanese

Permissions

All chapters in this book were first published in LDREAELTSFNA, by Springer; hereby published with permission under the Creative Commons Attribution License or equivalent. Every chapter published in this book has been scrutinized by our experts. Their significance has been extensively debated. The topics covered herein carry significant information for a comprehensive understanding. They may even be implemented as practical applications or may be referred to as a beginning point for further studies.

The contributors of this book come from diverse backgrounds, making this book a truly international effort. We would like to thank all the contributing authors for lending their expertise to make the book truly unique. They have played a crucial role in the development of this book. Without their invaluable contributions this book wouldn't have been possible. They have made vital efforts to compile up to date information on the varied aspects of this subject to make this book a valuable addition to the collection of many professionals and students.

This book was conceptualized with the vision of imparting up-to-date and integrated information in this field. To ensure the same, a matchless editorial board was set up. Every individual on the board went through rigorous rounds of assessment to prove their worth. After which they invested a large part of their time researching and compiling the most relevant data for our readers.

The editorial board has been involved in producing this book since its inception. They have spent rigorous hours researching and exploring the diverse topics which have resulted in the successful publishing of this book. They have passed on their knowledge of decades through this book. To expedite this challenging task, the publisher supported the team at every step. A small team of assistant editors was also appointed to further simplify the editing procedure and attain best results for the readers.

Apart from the editorial board, the designing team has also invested a significant amount of their time in understanding the subject and creating the most relevant covers. They scrutinized every image to scout for the most suitable representation of the subject and create an appropriate cover for the book.

The publishing team has been an ardent support to the editorial, designing and production team. Their endless efforts to recruit the best for this project, has resulted in the accomplishment of this book. They are a veteran in the field of academics and their pool of knowledge is as vast as their experience in printing. Their expertise and guidance has proved useful at every step. Their uncompromising quality standards have made this book an exceptional effort. Their encouragement from time to time has been an inspiration for everyone.

The publisher and the editorial board hope that this book will prove to be a valuable piece of knowledge for students, practitioners and scholars across the globe.

Index

A

Acrosomal Morphology, 121, 126

Apodemus Speciosus, 12, 16-18, 23, 28-29, 133

Arthropods, 41-45, 48, 50

Artificial Insemination, 121, 124

Atomic-bomb, 2, 170, 223

B

Biological Effect, 3-4, 58, 148-149, 153, 161, 173, 206, 229-231

Biomass, 42, 48, 63, 66, 69, 76, 78

Blood Plasma, 121-125, 132, 142, 212, 217, 219

C

Capillary Electrophoresis, 122-123, 125, 134

Chronic Inflammation, 136-137, 146

Coastal Area, 17, 39, 89

Coniferous Trees, 97-105

Conversion Factor, 19, 172, 175-180, 220

D

Does Rate, 17

E

Ecosystems, 13, 42, 49-50, 82, 88

Embryo Transfer, 110, 113, 115-116

Endothelial Cells, 2, 202, 205

Evacuation Zone, 1, 6-8, 12, 51-54, 56-57, 59, 97, 102-103, 110, 112, 119-124, 132-133, 136-141, 143-146, 148-149, 151-152, 155, 159, 163, 183, 206-207, 210

Ex-evacuation Zone, 7-8, 12, 51-54, 56-57, 97, 102-103, 110, 112, 119, 121-124, 132-133, 136-141, 143-146, 148-149, 151-152, 155, 159, 163, 206-207, 210

External Exposure, 2-4, 6, 24, 122-123, 126, 169, 172-173, 175-180, 183, 207, 210, 220-221, 230-232

F

Female Fertilities, 110, 112

Field Mouse, 17-20, 23, 27, 215

Fishery Products, 30-35, 37-38

Fishing Industry, 30-31

Fukushima Daiichi Nuclear Power Plant, 1-2, 7, 12, 16-17, 28-30, 83, 85, 87-88, 95, 97, 102-103, 106-107, 203, 206, 210, 213, 224, 229, 238-239, 247

Fukushima Nuclear Accident, 12, 28, 39-40, 120, 133, 155, 171, 183, 194-195, 203, 210, 236, 246-247

Fukushima Prefecture, 6, 16-19, 30-31, 35, 38, 42, 53, 56, 64-68, 77-82, 85, 89, 94, 102, 119, 123, 133, 138, 143, 161, 179, 181, 188-190, 192, 195

G

Gene Expression, 12, 136, 138, 140-141, 143-146, 201-202, 205, 234

Genetic Effects, 88, 110, 112, 223

Genetic Predictors, 87, 89

Granulosa Cells, 110-111, 113-114, 117, 119

Grasshopper, 41-43, 46

H

Histological Analysis, 125, 138

Human Health, 1-2, 6, 11, 197, 199, 206

I

Immune Responses, 136-137, 140-141, 144

Immune System, 135-137, 144-145

Inflammatory Cytokine, 136

Insoluble Radioactive Particles, 188, 197, 200

Internal Exposure, 2-3, 8, 20, 27, 58, 126, 169, 171-173, 175-178, 183, 206-207, 212, 214-217, 219-221, 230-233

Intertidal Biota, 63, 66, 78, 82, 85

Intertidal Invertebrates, 66-67, 73, 82

Intertidal Zone, 63, 66-69, 73, 76, 78-80

J

Japanese Macaque, 1, 173, 175, 178

L

Large Japanese Field, 8, 12, 16-18, 23-24, 28-29, 133

Liver Cancer, 2-3

Low-dose Radiation, 12, 29, 82, 112, 133, 146, 205-206, 212, 225, 236

Low-dose-rate, 110-112, 121, 123, 144, 146, 148, 152, 161, 202, 210, 212, 224

M

Masu Salmon, 37, 87, 89-90, 92, 94, 96

Mitochondrial Dna, 87, 89, 92, 96

Monitoring Research, 30, 32-34, 36-38

Muridae Species, 16, 18, 27

Mutation, 27, 87-89, 92-96, 153, 206, 208-209, 211, 214, 219, 222-226

N

Nonhuman Biota, 65, 97

Nuclear Accident, 12, 28, 39-40, 59, 66, 78, 83, 94, 111, 120, 133, 155, 159, 169, 171, 173, 183, 187-188, 194-195, 203, 210, 230, 236, 246-247

Nuclear Disaster, 39, 63, 84-85, 159, 161, 183-184

Nuclear Power Plant, 1-3, 7, 63-64, 67, 73, 83, 85, 87-88, 95, 97-98, 133, 135, 137, 146, 148, 154-155, 203, 206, 210, 212-214, 224, 229, 238-239, 247

Nuclear Power Plant Accident, 1, 7, 12, 14, 28-29, 38-41, 49-50, 52, 59, 83, 88, 95, 120, 122, 133, 154-155, 183-184, 194-195, 197, 206, 210, 212, 224, 229, 238, 247

O

Oncorhynchus Masou, 37, 87, 89

P

Pathological Changes, 2, 145

Phosphorylated Histone, 148-149

Plasma Biochemistry, 136, 142, 145

Population Decline, 63

Population Densities, 63, 66, 69, 73, 76-80, 82

Population Density, 63, 68-69, 76, 238, 243

Power Plant Accident, 1, 7, 12, 14, 28-29, 38-41, 49-50, 52, 59, 83, 88, 95, 120, 122, 133, 154-155, 172-173, 183-184, 194-195, 197, 206, 210, 212, 224, 229, 237-238, 247

R

Radiation Effect, 9, 97-98, 101, 161

Radiation Exposure, 1-3, 6, 12, 16-17, 29, 82-83, 99, 136-137, 140, 145, 148-156, 159-161, 170, 202, 207, 209-210, 212-214, 220-222, 224-225, 234, 245-247

Radiation-induced Cancer, 1, 153

Radioactive Cesium, 7, 10, 28, 30-32, 34, 39, 41-42, 50, 85, 87, 94, 111, 136-137, 184, 187-188, 195, 197-198, 210, 212, 215

Radioactive Contamination, 42, 50, 60, 95, 122, 137, 171, 238-240, 243

Radioactive Cs-bearing Particles, 187-194, 197-203

Radioactive Plume, 17

Radioactive Pollution, 31-32, 89, 155

Radionuclide, 32, 40, 43, 59, 64-65, 80-82, 97, 104-105, 107, 120, 123, 133, 175-176, 178, 180-181, 188, 238, 240-242, 244-246

Radionuclide Contamination, 43, 64, 97, 104, 107

Radiosensitive Plants, 97

Resonance Dosimetry, 161-162, 170

Rock Shell, 63, 66, 77-78, 82

S

Sessile Organisms, 63, 67-68, 76, 78-80, 82

Species Richness, 63, 78, 82

T

Thorotrast-induced Liver, 2-3, 10-11

Tohoku Earthquake, 63-64, 84

Tooth Dosimetry, 159-163, 165-171

Transgenerational Effects, 10, 121-123, 132, 212, 217, 236

Trial Fishing Operation, 30-31, 35-38

V

Viable Oocytes, 110, 118

Printed in the USA
CPSIA information can be obtained
at www.ICGtesting.com
JSHW010843251023
50683JS00016B/25

9 781647 403539